STAINS ON THE SCREEN
The Geometric Imaginary
and Its Contaminative Process

RESEARCH IN DESIGN
THEORY AND METHODOLOGY
A Selective Overview of
Recent Work in English

RESEARCH IN PHILOSOPHY AND TECHNOLOGY, Supplement 4
Series Editor: Carl Mitcham, Pennsylvania State University

Research in Philosophy and Technology is a refereed publication. Editorial correspondence should be addressed to Carl Mitcham, General Editor, Philosophy Department, Sparks 240, Pennsylvania State University, University Park, PA, 16802.

STAINS ON THE SCREEN
The Geometric Imaginary
and Its Contaminative Process
by NOËL GRAY

RESEARCH IN DESIGN
THEORY AND METHODOLOGY
A Selective Overview of
Recent Work in English
by GERHARD BANSE

Emerald

United Kingdom – North America – Japan
India – Malaysia – China

Emerald Group Publishing Limited
Howard House, Wagon Lane, Bingley BD16 1WA, UK

British Library Cataloguing in Publication Data
A catalogue record for this book is available from the British Library

ISBN: 978-0-76-230607-7
ISSN: 0161-7249 (Series)

Printed and bound by CPI Group (UK) Ltd, Croydon, CR0 4YY

CONTENTS

FOREWORD

The present volume constitutes the fourth special supplement in the *Research in Philosophy and Technology* series. It extends the *RPT* monograph tradition that began with Paul Levinson's *Mind at Large: Knowing in the Technological Age* (Supplement 2, 1988) and continued with Leonard Waks' *Technology's School: The Challenge to Philosophy* (Supplement 3, 1995). It also augments the *RPT* bibliography heritage exemplified by Joyce Hanks' *Jacques Ellul: A Comprehensive Bibliography* (Supplement 1, 1984), a legacy that in fact has roots in the "Current Bibliography in the Philosophy of Technology: 1972-1974" (in *RPT*, vol. 1, 1978) and "Current Bibliography in the Philosophy of Technology: 1975-1976" (*RPT*, full vol. 4, 1981).

STAINS ON THE SCREEN

Representing the first tradition is Noël Gray's monograph *Stains on the Screen: The Geometric Imaginary and Its Contaminative Processes.* As was the case with the two previous books in the series, Gray's volume undertakes an extended examination of an virtually unexplored issue in the philosophy of technology. Here, the question is not artifactual epistemology (Levinson) or the school-technology nexus (Waks) but the technique of projection that found its first clear expression in the geometric imagination.

There exists no comparable study in the philosophy of geometry and its techniques. The movement from the geometers Proclus and Euclid (chapter 2) to David Brewster and Benoit Mandelbrot (chapters 6 and 7)—with extended critical incorporation of related reflections by Michel Serres, Edmund Husserl, and Cornelius Castoriadis (chapters 3 through 5)—is a tour de force. It develops a thesis regarding a dialectic of purity and contamination: the screen or plane surface is never the pure or neutral mediation that has often been claimed. As contaminated by what is thrown or projected onto it, it also contaminates in its different incarnations as ancient (Euclidean), modern (Cartesian), and postmodern (Mandelbrotian) screen.

Given the contemporary technological incarnation of the screen in the computer, the implications for the philosophy of technology are significant—although they are left largely unstated by the author. In a personal communication, Gray has nevertheless ventured that he has come to understand that his work is a "rich field for critical investigation" because "a sense of geometry in one form or another is all but essential to the very possibility of the construction of technology and instrumentation." In saying this, Gray invokes:

> a wider sense of geometry beyond the confines of the view which understands geometry as primarily an ever repeatable rigorous system that issues up identity or sameness again and again. [Geometry is] a theory of knowledge and a theory of existence and a generator of meaning, co-extensive to the narrower sense of geometry as solely a mathematical tool for the purpose of extracting and later applying the laws that we take to govern nature and material construction. Hence to subsequently critique geometry within the frame of this wider sense, is in a way I believe, to start one's critique of technology...at the very etymological base that announces the sense of this term. That is, to direct one's critique of the very *making of logic* at the same time as directing one's attention to the *logic of making* or *what logic makes*; ....In other words, it is to start one's critique at the level of the meaning systems that play a formative part in the theorization, construction and subsequent perception and assimilation of technology in society, as well as to go to the very foundation of scientific practice and theorization in general.[1]

As this passage may already suggest, the book is itself structured in a poetic and yet quite rigorous manner. (In some ways, it resembles the work of Gaston Bachelard more than typical Anglo-American philosophy.) The introduction lays out three interpretations of the function of the screen, then opens up into three major parts, each with two chapters functioning as something like a first and second musical movement, progressively exploring approaches to the screen in its differentiated historical manifestations. There is a lengthening of each part in both pages and number of subsections that reflects a corresponding deepening of issues.

This most unusual argument is developed by a most unusual scholar. Noël Gray was born in Sydney, Australia, in 1944. At age 19, he became a sculptor, later specializing in stone carving, and over the next three decades he established a substantial reputation in the arts for the quality and vision of his work. His gallery work was featured in many group and solo exhibitions in Australia, Brazil, England, and Japan, and his commissioned work may be seen in Sydney, Liverpool, and Rio de Janeiro. Gray's drawings and paintings have also been featured in a number of exhibitions.

Perhaps the dominant feature of Gray's sculpture is its geometric and intellectual character. Visually, his large sandstone figures remind one of the Easter Island megaliths and the Mayan art of Central America, but with a more intricate mathematical structure. As Geoffrey Batchen wrote in *Forma Pura: A obra do escultor australiano Noël Gray* [Pure Form: On the Work of the Australian Sculptor Noël Gray], the catalog for a 1982 Brazilian exhibition of his work:

> technical and ingenuity and skill are only secondary aspects of the work of Gray....For Gray is notable among contemporary artists for his intellectual as well as his technical abilities. From an early age he has read extensively on philosophy and theoretical science, and his knowledge in these areas is a guiding influence on his work. Indeed the particular direction his sculpture has taken owes a great deal to his desire to be able to discuss these areas through the medium of art. It also explains the pre-occupation of his present work with the principles and practices of form.[2]

Following his intellectual interests, in 1985 Gray undertook a course of philosophical and fine arts study at the University of Sydney. Having earned the BA with honors and university medal in 1989, he went on to complete the PhD in 1994, after which he was awarded traveling fellowships by the Eleanor Sophie Wood Trust of the University of Sydney and by the Australian government. With fellowship support, he undertook further research in the United States and Europe, where he completed the present book. He currently lives in Italy, where he continues to write.

It is with pleasure that *RPT* publishes his original study of what may be called the central technique (or technology) of geometry, an area of mathematics that has in turn been of major influence on modern technological thought and practice—as well as on attempts to think technology philosophically. Gray's study has the potential ability to contribute to a radical reappraisal of both geometry and its technological influences. Certainly, his notion of a dialectic between purity and impurity, and a contaminative process present in the very heart of the geometric attempt to purify experience, is a phenomenon characteristic of modern technology as well.

DESIGN BIBLIOGRAPHY

Representing the second heritage is Gerhard Banse's selective bibliography on "Research in Design Theory and Methodology." As has been the case with previous *Research in Philosophy and Technology* bibliographies, Banse undertakes a literature review of special use to the philosophy and technology studies field.

Moreover, although not quite as unexplored an area as Gray's question of the geometric imaginary, the issue of design is surely a neglected but significant field for future philosophical reflection—and one not wholly unrelated to that opened up by Gray. Indeed, it was because of this relation, however unacknowledged by the two authors, that it was judged appropriate to include the monograph and bibliography in a single supplementary volume.

Design is an activity at the center of that dimension of technology known as engineering, which itself must be granted a prominent place in any appreciation of modern technology. Despite this scarcely disputable fact, design and the designing process have been accorded only the most cursory attention in the philosophy of technology. Banse's bibliography is designed to stimulate the kind of interest and reflection that may begin to overcome such indifference.

Banse himself is a philosopher and social scientist from the former German Democratic Republic, and even before reunification, he had become a contributor to the philosophy of technology outside the confines of rigid ideological Marxist positions.[3] Since German reunification, he has joined the faculty of Technische Universität Cottbus and has worked especially to develop an interdisciplinary dialogue with the technical community.[4] This bibliography constitutes an effort to initiate assessment of English research in an area of particular importance in a way that may at once advance interdisciplinary German work in the technical and human sciences, build bridges between German and English research programs, and contribute to an existing English language research discourse.

It may also be noted that Banse's work exists in part within the institutional context of a unique Lehrstuhl für Technikphilosophie at the Technische Universität Cottbus, that is, a Chair in the Philosophy of Technology held by Klaus Kornwachs, who is himself an exceptional proponent and practitioner of interdisciplinary research.[5] With degrees in both the sciences and philosophy, and having held industrial positions both in Europe and abroad, Kornwachs is in a singular position to appreciate the need for, to exercise leadership in, and to sponsor a research team in interdisciplinary philosophy of technology, to which Banse has contributed. Indeed, it is within such general interdisciplinary programs that the monograph by Noël Gray on the geometric imagination and the bibliography by Gerhard Banse on design theory most properly find their homes.

Carl Mitcham
Philosophy Department and Science,
Technology, and Society Program

NOTES

1. Personal communication from Noël Gray, February 1995.

2. Untitled introduction, in Noël Gray and Geoff Batchen, *Forma Pura: A obra do escultor australiano Noël Gray* (Rio de Janeiro: Museu Nacional de Belas Artes, 1982), p. 10.

3. See, for example, Gerhard Banse, "Philosophische Fragen der technischen Wissenschaften—Probleme und Ergebnisse" [Philosophical Questions of the Technical Sciences—Problems and Results], *Deutsche Zeitschfrit für Philosophie*, vol. 24, no. 3 (1976), pp. 307-318; Gerhard Banse and Siegfried Wollgast, *Biographien bedeutender Techniker, Ingenieure und Technikwissenschaftler: Eine Sammlung von Biographien* [Biographies of Important Technicians, Engineers, and Technoscientific Workers: A Collection of Biographies] (Berlin: Wolk und Wissen Bolkseigener, 1983); and Gerhard Banse and Helge Wendt, *Erkenntnismethoden in den Technikwissenschaften: Eine methodologische Analyse und philosophische Diskussion der Erkenntnisprozesse in den Technikwissenschaften* [Epistemological Method in the Technological Sciences: A Methodological Analysis and Philosophical Discussion of Knowing the Technological Sciences] (Berlin: Verlag Technik, 1986).

4. See, for instance, Gerhard Banse, *Risiko—Technik—technisches Handeln: Eine Bestandsaufnahme* [Risk—Technology—Technical Action: A Resource Inventory] (Karlsruhe, Germany: Kernforschungszentrum Karlsruhe, 1993); Gerhard Banse, "Konstruieren im Spannungsfeld von Kunst und Wissenschaft: Historische Anmerkungen in systematischer Absicht" [Construction in Tension between Art and Science: Historical Notes in Systematic Perspective], *Technikgeschichte*, vol. 61, no. 4 (1994), pp. 329-352; Gerhard Banse, ed., *Risikoforschung zwischen Disziplinarität und Interdisziplinarität: Von der Illusion der Sicherheit zum Umgang mit Unsicherheit* [Risk Research between Disciplinarity and Interdisciplinarity: From the Illusion of Safety to Relations with Unsafety] (Berlin: Sigma Rainer Bohn, 1996); and Gerhard Banse and Käthe Friedrich, "Technische Welterzeugung—Philosophische Aspekte" [Creation of the Technological World—Philosophical Issues], in Christoph Hubig and Hans Poser, eds., *Cognitio humana—Dynamik des Wissens und der Werte*, Proceedings of XVII. Deutscher Kongreß für Philosophie, Leipzig, 1996, vol. 1 (Leipzig: Institut für Philosophie, 1996), pp. 334-341.

5. For representative work, see Klaus Kornwachs, ed., *Offenheit, Zeitlichkeit, Komplexität: Zur Theorie der Offenen Systeme* [Openness, Temporality, Complexity: On Theories of Open Systems] (Frankfurt am Main: Campus, 1984); Hans Jorg Bullinger, B. Shackel, and Klaus Kornwachs, eds., *Human-Computer Interations: INTERACT '87*, Proceedings of the Second IFIP Conference on Human-Computer Interactions, Stuttgart, September, 1987 (New York: Elsevier, 1987); Klaus Kornwachs, *Information und Kommunikation: Zur menschengerechten Technikgestaltung* [Information and Communication: On the Human Aspects of Technological Development] (New York: Springer, 1993); Klaus Kornwachs and Konstantin Jacoby, eds., *Information: New Questions to a Multidisciplinary Concept* (New York: John Wiley, 1996); and Ernst U. von Weizsäcker, Ivan Illich, and Klaus Kornwachs, *Teilen* [Sharing] (Ulm: Humboldt Studienzentrum, 1996).

Illustration 1. Frontpiece from David Gregory's edition of Euclid's *Opera* (Oxford: 1703) illustrating the shipwreck of Aristippus as related in Vitruvius, *De architectura* VI, Preface: "Aristippus the Socratic philosopher, having been shipwrecked on the coast of Rhodes, observing geometrical schemata drawn [in the sand], is said to have exclaimed to his companions: 'We can hope for the best! I see the traces of humans'." (By permission of the Houghton Library, Harvard University.)

STAINS ON THE SCREEN:
The Geometric Imaginary and Its Contaminative Processes

by NOËL GRAY

TO VERONIQUE FOTI

ACKNOWLEDGMENTS

The author wishes to give special recognition and appreciation to Christine R. Bailey of the Rare Books Room at the Pennsylvania State University Libraries for her exceptional work in helping to locate illustrations useful for enlivening this text. Others to be commended are the editor and his more than able assistant James Frazier.

***Illustration* 2.** "Antrum Platonicum" (1604), an engraving by Jan Saenredam (1565-1607). See the "Note on the Illustration of Plato's Allegory of the Cave" at the end of the text.

Chapter 1

INTRODUCTION

PLANES AND SCREENS

The phenomenon of the screen, in its ancient form as a two-dimensional flat plane, is at the heart of the long history of geometry.[1] It is constituted as a site awaiting or inviting projections, reflections, marks, shadows—a site onto which images are either cast or written, reflected from, or spontaneously generated within. Whether as an empirical phenomenon or a transcendental one, its essence has been that of a device standing between realities. As an intermediary form, it acts to reveal in ways favoring one aspect of reality over another. It also acts to overcome what is perceived as the limits of a specific reality.

For instance, Thales determined the height of the great pyramid by measuring the shadow it cast on the sand, thereby making the inaccessible (out of physical reach) accessible to measurement. The sand acted as a screen standing between the geometer and the pyramid. It functioned to overcome the physical limitations of the geometer. The privileged reality, the true height of the pyramid, was thus brought down to earth—to the lower, accessible world of the geometer; relocated, it subsequently imbued this lower reality with its higher status.

Plato also used this idea of the sand as a screen or plane which revealed the inaccessible. In the *Meno*, Socrates drew geometric figures in the sand as a device to elicit inherent geometrical knowledge from a slave, knowledge that the slave did not know he possessed. By aiding recollection, the diagram on a plane subsequently became installed as an ancilla to the attainment of true knowledge—a situation that continues to this day in the form of the common building plan stored by local planning authorities. In the *Republic*, Plato further employed the plane as a revelatory device in his simile of the cave. The form it takes is of a pool at the mouth of the cave. This pool, in being written upon by light from the heavens, reflects upon its planar surface geometric figures such as circles (planets) and points (stars).

Because of the brightness of the truth (the sun) the pool acts as both reflective plane and darkening glass which discloses the truth at the same time as it protects the freed prisoner against the full glare of this truth/sun. Such preparation is needed because his previous existence has been lived in a world of shadows. Shadows he took to be true reality. The images in the pool, austere geometric figures, are thus installed in the history of philosophy and science as that which reflect, approximately, the simplicity and static character of Platonic ideas.

Euclid also incorporated the phenomenon of the plane but granted it a certain etherealness—the form of an endless surface which the geometer imagined as the (ideal) site upon which the main elements of geometry carried out their functions (ideally). Indeed, Euclid's endless plane was as ideal as that which it served to illustrate. Thus, in this early history of geometry, the plane takes on the status of an ideality (although, for Plato, this plane, precisely because it was imaged, did not have this status). Alternatively, for Euclid, the endless character of its surface overcame the inherent sense of limit associated with any image.

Later in this history comes the Cartesian two-dimensional grid, upon which points in three-dimensional space can be projected and plotted. This new screen allows for a schematization in advance and a prediction of events in the physical world yet to be enacted. The thickness of the world, so to speak, where everything is in other words, becomes conveyed in advance to anyone desiring such information. One can also tabulate, reckon, give a number (set of coordinates) to anything in perceivable or imagined space. Such a grid can even erase itself by being transformed into a set of numbers.

The same period saw the emergence of Alberti's transparent screen, which was placed between an observer and some scene in need of being rendered. Upon this particular screen's surface, straight line projections traced an image, making it possible for two-dimensional or planar representations to (look like or) correspond, point by point, with the three-dimensionality visible to ordinary perception. This germinal device overcame the limits of previous pictorial renderings: their lack of perspective or apparent lack of agreement with ordinary perception. In more recent times, we have the ubiquitous blackboard, the transparencies of overhead projectors, and the computer screen. All these devices stand, in one way or another, between the viewer and some geometric idea in need of communication. The first two can be written upon and erased at will, while the last can be made to be self-generating and self-erasing.

This is a spare list. But, what binds these phenomena together is the fact each plane or screen is traditionally thought to be neutral, to be merely a convenient device or vehicle for the communication of some idea. In contrast to this tradition, then, I will argue the screen becomes intricately interwoven with both the structure and the construction of what is in need of communication. Which is to say, in another way: the geometric imaginary, and whatever screenic-planar phenomenon is mobilized to convey or illustrate its nature, are both deeply enfolded within each other. One cannot think the one without thinking the other. But, for the moment, let us nevertheless assume their separation.

The foundational etymology of the term "screen" is clouded, but the word has a loose connection to Old French (écrin, escrin: *jewel case*; and possibly to esconse: *hiding place*; and possibly to écu, escu: *shield*), and remotely to Latin (scrinium: *a container for papers and books, a writing desk*; scutum: *shield*; velum: *a sail, covering, awning, curtain*), and even more distantly to Dutch (schans, schantze, schentze: *screen of brushwood for soldiers*). Current usage has retained and expanded on this sense of *to hide*.

"Screen" now incorporates a very complicated set of meanings that have a wide range of applications beyond the idea of concealment and containment. This range of meanings can be grouped under three dominant functions: to cover, reveal, and interpose/transpose; and under three dominant properties: opacity, transparency, and reflection. Such functions and properties get used in pairs or in a variety of combinations or mixes, regardless of whether the screen is a material or ethereal form or process.

In the first function, covering, the screen serves to hide, conceal, deceive, and protect. This function often relies on the property of opacity in both the spectral and conceptual senses: a barrier to light, observation, and natural forces, or to thought, comprehension, analysis, memory, or recollection. But even a transparent screen can sometimes hide, conceal, deceive, or protect. A transparent glass, for instance, can conceal the way it might transform rays of light, or even its own material presence. It can also protect (as a windshield does), by deflecting harmful rays of some type. A screen that is reflective can hide what is behind its surface. It can deceive by multiplying instances or by problematizing identity. Screens can also cover by becoming surfaces filled with arresting or absorbing images, as with movie and television screens.

In the second instance, revealing, the screen discloses by some process, construction, or transposition. Such revealing often relies on the property of transparency or a "seeing through." For instance, as employed by Renaissance artists, a lined screen set up before a scene superimposes a grid on it, enabling a rigidly held eye to guide the hand for constructing a perspectival image. This reveals, orders, or formulates the geometric character of the scene being observed and, thus, produces a (geometric) correspondence between the physical object observed and its graphic representation.

Another example of the revelatory screen is shown in the form of a photographic negative which, when placed between a light source and a chemically sensitive plate or paper, reveals the positivity of the original subject image. The latter is held to correspond with ordinary observation. Something related occurs with the silk screen when it transposes its image to any number of other appropriate surfaces. However, an opaque screen can also reveal—as in the idea, attractive to many empiricist philosophers, that visual sense data become projected, through the retinal screen, onto the mind as an unmediated truth, or a truth directly perceived.

The reflective screen can also reveal, in the same way as a television or movie screen reveals some element of reality in a concentrated or novel way.

Finally, as something interposed, the screen functions as a regulatory form or process that divides, separates, selects, and distorts. This interposition, being placed between or in front of, often relies on the property of reflection. The cinematic screen reflects a set of projected images of a highly orchestrated and selective reality that can distort everyday experience. We also have reflector screens that regulate light and harmful rays, or intensify them, as in suntanning reflectors and solar ovens. Moreover, the previously mentioned opaque and transparent screens can, when interposing, obviously serve also to divide and separate. A sieve or grading screen—which, as a spatial grid or net, is a kind of physical and visual transparency—can regulate by sorting and selecting. And, in a completely unique fashion, the x-ray plate is a screen that receives and reveals the skeletal images of an irradiated object.

In short, even from these brief descriptions of the screen as covering, revealing, and interposing—descriptions which in no way exhaust the manifestations of such histories—the screen acts as a device or process that mediates, frames, gathers, produces, hides, disperses, or differentiates one thing from another. It separates one reality from another, or reveals one in another. It is also apparent that the screen can be both a material site and an ethereal interface upon which different realities may meet or upon which the cohabitation of different realities is possible. However, beyond any metaphor of separation, cohabitation, or neutrality, the screen also contaminates both sides of that which it covers, reveals, or transposes. Specifically, at the level of the geometric imaginary, it is a staining or contaminative process at the same time as it functions to cover, reveal, or interpose/transpose.

GEOMETRIC IMAGINARIES

Putting aside this idea of contamination until later, one thing to be noted is the historically repeated reformulation and reconstruction of the geometrical imaginary as a plane or screen upon which geometry displays and manifests its theoretical utterances concerning space. For example, Proclus, in contrast to the modern idea that the geometric imaginary deals solely in abstractions, argues that his planar geometric imaginary is less pure and self-same than what he calls the upper reaches of intellection or pure thought. Yet, this plane is at the same moment also more pure than the lower world of the sensible.

Developing this Proclian idea, Michel Serres introduces the idea of cacography (the irregularity of the graphic figure). He argues that overlooking such irregularity is the way in which communication—say, geometric abstraction in the form of idealities—becomes possible.

In another historical reconstruction of the geometric imaginary, Edmund Husserl argues that geometry has become, in its purity or abstraction, divorced from the vicissitudes of the life-world, divorced from the becoming or creative character of the abundant world of the experiential. Extending Husserl, cultural theorist Cornelius Castoriadis introduces the idea of abstract space and the possibility of it being emptied of its contents.

In yet another historical variation, David Brewster's kaleidoscope, with its internal reflective screens or mirrors, emerges as a materialized geometric imaginary that gathers purity and impurity in a way that challenges traditional geometries. Ironically, Brewster's instrumental critique forced the kaleidoscope to be excluded from the history of serious scientific instruments. Finally, in more recent times, Benoit Mandelbrot has offered his fractal images as another erasure of the traditional geometric imaginary. Mandelbrot's computer screen emerges as an electronically generated geometric imaginary reproducing on its own electronic surface both mental and perceptual images.

What will later become apparent in the extended studies of these historical moments, is that once geometry takes shape and the binary character of its theoretical structure becomes clear, then it is possible to think of geometry as a discourse that is as much about separation as about unification—as much about theorizing spacings as theorizing the totality of space. Specifically, the divided terrain that the geometric imaginary invariantly constructs on a screen to privilege some notion of a pure (or purer) reality over an impure one, is itself constantly traversed or contaminated by a certain impurity.

Indeed, like the footprints of some lightweight dancer on a glacier, it is possible to find traces of this impurity in the myriad of developments of Western science from ancient Greece to the present day. In fact, this trail of impurity can be found as much as any other feature of science, even as much as the oft-stated unquenchable thirst for pure, objective, or novel truth. However, so faint is this trace or stain, so enfolded or enmeshed is it in the grounding discourse of all science (geometry), that there is little wonder its significance in the privileging processes of science has been all but overlooked. To explicate, therefore, the significance of this stain, and because of the paucity of theorization of the geometric imaginary itself—and because my intention is to move beyond the dominant voice of geometry—it will be necessary in a genealogical fashion to interrogate the margins of science. We must look toward the edges, so to say, to glimpse the deeper character of this science of edges.

Understandably, neglect of the geometric imaginary has been encouraged since the time of Galileo by the promotion of geometry as a tool for the discovery and rigorous description of space and as a program for materializing or constructing, mathematical conceptions.[2] But, this modern interest in the explanatory and constructive character of geometry has a forerunner in the Platonic exemplar: the inscription over the entrance to Plato's academy extolling all who entered to be conversant with geometry. This commandment gave to geometry the status of a Masonic law for science, something to be not investigated but only followed. Thinking in the fashion of a geometer became the primary example of how to grasp the unchanging, universal elements or transcendent features of existence. This commandment further set in motion the conditions for understanding reality as a constant tension or iterative dance between the ideal and the empirical—between the abstract and the applied, the static and the dynamic, the ordered and the chaotic.

So powerful was this exemplar that geometry became, up to and including Immanuel Kant, the one clear certainty at the same time as it was a tool for the construction and understanding of mental and physical worlds. Although the post-Kantian appearance of non-Euclidean geometry fundamentally altered this idea of unassailable certainty, what remained was the idea of geometry as the truth of space. Geometry quâ geometry. Geometry remained thought of as an impartial system, one that continued to cover, reveal, and transpose a set of repeatable procedures—procedures, moreover, which were capable of bringing the future into the present in the form of predictive statements. One could see before one had physically made it, what a thing would eventually look like, or how an event would turn out. In the wake of this extraordinary feature, geometry—and this was true of both non-Euclidean and Euclidean versions—thus became the science that formulated a relation between permanence, construction, and prediction.

One culmination of this kind of prediction was the first successful landing of a human being on the moon in 1969. But for geometry quâ geometry, the trip to the moon was not the first time it had enabled humans to rise above the earth. Such transposition of the human can be found further developed in the architecture of the computer (an essential apparatus in space flight), which also owed much to geometry. Indeed, the modern world. with its transport systems, computers, skyscrapers, weather and financial forecasts, and almost anything else one might care to imagine, would be to most of us inconceivable without the endless and certain repeatability of geometry and its constructive and predictive powers.

In light, then, of this general or prevalent view, and mindful of geometry's undeniable and unparalleled successes at the level of the everyday, there is little wonder that the imaginary, with all its apparent ethereality and less-than-objective associations, might appear as somewhat beside the point. If not wholly beside the point, then surely of secondary import to geometers and users of geometry alike. Rather, for them, the concern has been primarily with developing new geometries or discussing the philosophical claims of some geometric theory and the ramifications thereof—just as many historians of geometry have been mainly concerned with formulating models which argue the historical changes in geometry as fundamentally evolutionary, or discontinuous.

But now, by concentrating as well on the imaginary and its processes, one can also show that geometry (and perhaps science in general) has deeply enfolded within it a constant need to speak of impurity, first, as a way to privilege purity/regularity, and then, as a way to overthrow it. In this manner, the study of geometry can thus be expanded into a way to more fully comprehend how we understand

and construct the distinctive features of our individuated, ever changing, yet shared worlds.

THE VEIL OF REPRESENTATION

In speaking of different worlds, the development and proliferation in the latter half of this century of high speed computation, high resolution graphics, and computer simulation has lead to the vision of a denser and more extensive experience of reality by newer technologies. In one of its most recent manifestations, this vision is offered through a new, or virtual, reality, freed from the vicissitudinous and monotonous character of everyday life. At the same time, in professional literature, seminars, and conferences, and even in the popular media, we also find contemporary science hailing fractal geometry, chaos theory, self-organizing criticality (or, simply, the theorization of the processes of change and emergence—that is, creation) as a new paradigm freeing science from the yoke of a past reductivist program grounded in geometrical idealities. But the combination of these highly visible forms of electronic technology and the complexity paradigm has produced a corresponding need in the humanities to understand more fully the processes by which, and how, science and technology produce the meanings upon which many of their theories and predictive claims are founded. This desire for a fuller understanding can have two foundations.

First, it can grow out of a need to establish an increasingly informed and critical base from which sociological, political, moral and ethical perspectives can engage technological and scientific developments (rather than simply be engulfed by them). Along these lines, one way to achieve a fuller understanding is by advancing an initial perspective that is itself prescriptive, prophetic or predictive—informing the community of the possible ramifications of existing and projected science and technology. Another would be to advance a critical attitude that is postscriptive or legislative—encouraging governments to enact legislation either to contain or eradicate a problem generated by technology or to ensure that a particular problem is not repeated. Ecological and environmental activists, lobbyist groups, and educative institutes and their professionals have successfully employed both stratagems. And their message is clear: science is too important to be left solely to scientists, no matter how substantive and concerned are the latter's operational and theoretical intentions. As it has often been said, this planet (and its inhabitants) is not a laboratory subject or site wherein and on experiments may be conducted with impunity.

Second, the desire for fuller understanding can emerge in an aspiration to intervene in the initial formation of scientific theories. For instance, the recent promise of a new generated reality through a science of processes also contains a clear announcment that *how* the experiential is thought (including its mathematization) is central to the deployment and development of postmodern natural science. This recent turn of events is not without irony, since the undeniable past success of science in defining its own practice as the discoverer of the unchanging truths of existence, also meant that it had to acknowledge by default that the individuated experiences of existence fell to the domain of the humanities.[3] The initial success of this pairing off of intellectual tasks is readily witnessed in the writings of scientists about various subjects now given over to the humanities, such as mythology, politics, and the realm of subjectivity. At its best, this writing still tends to lack a firm grasp of the current theoretical work being carried out in other disciplines. It tends to be missing the rigor that science demands of its own sphere and that it requires of the humanities when they venture to speak of science.

To therefore avoid, in its own turn, this pitfall, part of the interest of the humanities in a fuller understanding of the new world of a science of electronically figured complexity must be to appreciate more clearly the interest of contemporary science in the dynamic complexity of the experiential. This interest by science is in the first instance a disavowal of its traditional idea that experiential complexity is the veil of nature behind which nature hides its laws—we now live within the scientific promise of seeing nature's face in all its *nakedness*.

This promise appears fated to shock many philosophical pretensions, since not focusing on but going behind this mercurial veil has been the primary aspiration of Western science. Recall, again, this aspiration as announced in the moment of revelation outside Plato's cave of representation. In escaping from the lower reality of shadows and representations, the prisoner gazes upon the geometric-like reflections in the cool pool of water beneath the blinding noonday sun. In so doing, he, as it were, imbibes the austere geometric figure as a means to slate his thirst for the truth of nature quâ the cosmos. In this, Greek geometry is lead to a kind of intoxication with cosmology—a maieutic system of self-proclaimed mediation that raises the mind from earth to the heavens and measures the latter's timeless truths. In thus partaking of the geometricals, we lift the *veil of representation*—we get to see behind the screen that hides.

To continue for the moment this slightly altered perspective of the one I advanced at the outset of this introduction, we can note, as well, two other moments in this history. First, there is Galileo and Newton's static nature of non-relative principles, which also cast its own veil over nature and informed the unparalleled success of modern science by reenforcing the old Greek idea of unchanging first principles. Then, there is Descartes' fireside reverie that further installed the mind as the site of calculated abstractions, itself informed by a partial debt to Proclus' mathematical screen of the imagination. Proclus located this screen or plane in the Platonic schema between intellection and perception, thereby overcoming the dilemma posed by this schema regarding where to situate the austere graphic forms of the geometricals. That is, for the geometricals to be Platonic exemplars, they can not be derived initially from the lower world of the empirical, for this (as Proclus consistently tells us) is a deficient world lacking in discursive precision. Equally, they can have no place as images in the Platonic *nous*, for this realm partakes of no division. One might, therefore, marvel at Proclus' wit in dividing yet further Plato's divided reality by splitting mentality into pure intellection on the one hand, and the plane of the creative geometrical imagination on the other; the latter being reflective of the former but self-generative.

One might also, within the tradition of this Proclian desire for geometric idealization, further admire Descartes for erasing almost entirely the need to shackle geometry to graphic figures. I refer to algebraic notation, the reduction of the diagram to letters and numbers. We may conveniently see in this reduction into abstraction an impetus for Newton's calculus grounded in a displacement of perception: visible movement abstracted to an infinite number of minute, fixed numerical differences. Further, let us also note Kant's a priori of space that brought the veil down further, leaving ironically in its wake the inexhaustible desire to know the *thing-in-itself* as it-knows-itself: a knowing precluded, yet generated by the very logic of an a priori at the level of Kantian Subjectivity. Indeed, the always elusive yet always desirable a priori (and its ally, the non-mediate fundament of meaning) has so effectively haunted the discourse of science and philosophy that one may easily be forgiven for overlooking in its very ubiquity its lasting significance in the constant restructuring of the very conception of a divided reality, Platonic or otherwise.

Finally, in criticizing modern science, Husserl argued that since the time of Galileo, nature itself had been idealized and become a mathematical manifold. He called this the "mathematization of nature."[4] Discussing this insight, Charles Harvey noted that for Husserl, this meant that science became grounded in a metaphysics that ascribed an economy between *reality* and what is *most knowable*. Hence, what can be quantified or measured became equated with *true Being*, the truth of nature.[5] Because these principles are static and perfect and transcendent and, as such, could logically have no actual corporeality or existence as objects in space and time, then any expression of them must by definition constitute an approximation. Which is to say that an approximation, in either its theoretical or empirical modalities, is grounded in the metaphysics of the unrealizability of pure transcendent principles. Approximations became understood therefore as advancing slowly toward, but never arriving at, the ideal perfect truth of nature. This idea could also be expressed as the endless march of hypotheses aided in their journey by the power of predictive truth, the throwing-forward feature implicit in the idea of approximating an ideality. Thus the common scientific phrase "expressing-in-principle" is shorthand for the endless construction of approximations of ideal universal laws. The always elusive perfect form of these principles also meant that ultimate proof became forever delayed.

What this amounts to, if we follow Husserl, is that the practice of endless approximations is what in the past informed the possibility of modern science as an unending project. Furthermore, the equating of truth with measurement meant that this mathematical manifold achieved a higher ontological status than the life-world. The essential task, therefore, under this manifold became the formulation of this ideal truth against the backdrop of the abundant, chaotic world of appearances. However, as Harvey also notes, a paradox of method emerges from the idea that all of nature can be brought under the mathematical. The problem becomes the difficulty of quantifying the realm of the experiential while retaining the idea that lawful principles express the truth of the world. Clearly, nature cannot, at the same time, be subsumed under measurement yet also have an aspect beyond measurement. But equally, mathematical truth, in being thought of as the transcendent character of nature, must by definition be essentially separate in some manner from the apparent immeasurable and immediate life-world of the Subject.

Even Galileo's positing of the universe as a mathematical field or container of correspondence (as his attempt to circumvent this paradox might be expressed), wherein measurable changes in geometric properties are argued to correspond more or less to changes in the operation of the senses, only serves to introduce the idea of primary and secondary qualities. These former qualities are thought to be essential, and the latter might be called merely *effects* of the *ideal* reality. That is, secondary qualities emerge as indices of mathematical changes quâ the primary qualities of nature (Pythagoras' division of a musical string into different lengths comes to mind as a precursor to this idea). What we need to remember about this Galilean ploy is not so much that it is a weak solution to the above paradox but that it enshrines the idea that the real truth of nature is outside the Subjective (God as mathematician). Thus, the truth of nature became proposed as a set of universal quantifiable causes and hence, by extension, the (primary) ontological truth of the world is mathematical.

To stress this even more strongly: the Subject's life-world becomes itself measured by a relation to the metaphysics grounding modern science. This means that science contaminates how we are able to think about the life-world. Ironically, however, Husserl failed to take the latter step precisely because he did not fully note the contaminative and formative role of science in the very idea of a life-world. However, now, in combining the idea of metaphysical with physical contamination, we may profitably reread Husserl and build upon what he had in mind when he spoke of an ontological reversal. Recall that the life-world, for Husserl, far from being understood as initially grounding the practice of science, falls prey to an ontological reversal whereby Galileo's manifold of the *most knowable* becomes the *real* reality over the life-world. To extend this Husserlian insight, then, it only needs saying that to argue, as science has done in the immediate past, for nonrelative truths is also, necessarily, to postulate and displace a separate world of relative values (that is, the life-world). A world that, as I mentioned earlier, became by default (from the viewpoint of science) the province of the humanities.

But this tradition of the idealization of nature by mathematics has apparently faded away. We now live with a promise of the mathematization not of what lies behind nature's veil but of the veil itself. This promise signals a reversal of scientific interest from ideal principles to the immediate dynamics of the world. Far from conceiving nature as grounded in a fundament of unchanging laws, science now sets out to conceive nature as the expression of local processes that constantly interweave and jostle one another in a gigantic cosmic dance. Such processes are unable—indeed, unwilling—to ever keep to some ideal order. What distinguishes this new science from its past is that it now seeks to understand nature by replicating it through the employment of mathematical processes, rather than by discovering hidden laws expressed as mathematical abstractions. By extension, this reversal also marks a move of science toward placing experience—and we might even say culture—directly and expressly under the mantle of the mathematical.

However, in wanting to lay some foundation for a possible intervention at some later date in the theory formations of this new science, we would have to go much further than just noting the rise of a new era of empiricism. For instance, if we allow that contemporary science has an increasing interest in the irregularity or dynamic character of the world, then it is important to question to what degree the science of the immediate and ancient past became exclusively concerned at its deepest theoretical level with the world of idealities. We need to question to what degree the science of the past was exclusively about constructing a horizon of measurability from which its veridical gaze stared forth. Or again, in a reverse fashion, and recalling the first part of this introduction: What, in some of the basic theoretical structures of the science of the past, has been the role of irregularity, of impurity?

In attempting to engage this question, what will emerge in the following study is that geometry has been (and continues to be) as much involved in the theorization of a horizon of immeasurability as it has been (and continues to be) involved in the theorization of a horizon of measurability. To speak in the vernacular: science, in being thought of as the dance of iteration, requires two partners, both of which are always embracing and always jostling for the lead. It is this dance that we will now (re-)join.

NOTES

1. The term screen did not come into use until around the time of Alberti. Before this period, the geometric imaginary is spoken about as a plane, or planar surface. I am grateful to Carl Mitcham for alerting me to this, and to Veronique Foti for pointing out Morrow's poetic license regarding his translation of Proclus. There is no evidence of the term screen in any of the older translated works Morrow relied upon for his version. His use of the word to describe Proclus' geometric imaginary is no doubt occasioned by the fact that Proclus speaks of a "projection." However, since this study is not a history of the screen as such, the terms plane and screen are used interchangeably in text.

2. David Lachterman, *The Ethics of Geometry: A Genealogy of Modernity* (New York: Routledge, 1989). Also related: Patrick Heelan, *Space-Perception and the Philosophy of Science* (Berkeley, CA: University of California Press, 1983).

3. Edmund Husserl, *The Crisis of European Sciences and Transcendental Phenomenology: An Introduction to Phenomenological Philosophy*, trans. David Carr (Evanston, IL: Northwestern Press, 1970).

4. Husserl, *Crisis*, p. 23.

5. Charles Harvey, *Husserl's Phenomenology and the Foundations of Natural Science* (Athens, OH: Ohio University Press, 1989), pp. 13-18.

Illustration 3. Portrait of Fra Luca Pacioli (1445-1517) by Jacopo de Barbari (1440/50-1516), from the Museo Nazionale di Capodimonte, Naples, Italy. The Dominican Friar Luca Pacioli joined in his person the austere lives of religious and mathematician. His *Suma* of 1494 aimed to summarize all mathematical knowledge: arithmetic, algebra, geometry, and trigonometry. His Latin translation of Euclid's *Elements* of 1509 became the first printed edition of that work. (By permission of Alinari/ Art Resource, New York.)

PART I

THE AUSTERE SCREEN

Illustration 4. Illustrated title page of *The Elements of Euclid Explained*, "in a new and most easie method: together with the use of every proposition through all parts of the mathematicks. Written in French, by that excellent mathematician, F. Claud. Francis Milliet de Chales, of the Society of Jesus; now made English, and a multitude of errors corrected, which had escaped in the original." 3rd edition. Oxford: 1700. (Reproduced courtesy of the Rare Books Room, Pennsylvania State University Libraries.)

Chapter 2

PROCLUS' COMMENTARY ON EUCLID

INSTALLING THE SCREEN

Geometry is a discipline that appears more so than any other to span the dichotomies of science/technology and theory/practice. Associated equally with Euclid's deductive science and with the surveyor's technique, geometry appears to be a phenomenon which, although it became present in culture by means of a historical path, early transcended its origins even as it remains firmly present and influential in our world. In what follows, I want to suggest that this path of development is more complex than the standard evolutionary history of geometry presents. In fact, far from a solely linear development, what the following essay will point up is a geometry that zigzags between, as it constantly remains within, two poles.

While many aspects of its ancient heritage in surveying were formative elements in the origins of axiomatic geometry, the notion of the geometric imaginary received its first substantive theorization in the work of the Greek philosopher Proclus.[1] In his commentary on Euclid, he expands the Platonic schema of upper and lower realities by interposing between them the intermediate world of the geometric imagination.[2] Proclus speaks of this as a screen dividing the upper realm of *nous* (a type of pure intellection) from the lower realm of the sensible (appearances). This intervention gives a logical structure to the Platonic system by overcoming the difficulty of where mathematical entities (geometricals) are to be situated.

This difficulty is generated by the fact that in parts of Plato's work, there is an ambivalence regarding geometry using drawn figures. Properly speaking, drawn figures must belong to the lower world of the sensible. This poses a dilemma, since Plato also wishes to install geometry as exemplifying the best method to grasp the upper world of Ideas. The capacity for geometry to repeatedly recall the same figures and procedures thus becomes a metaphoric practice with which to grasp the static universality of idealities. Plato never adequately resolves this tension between relegation and metaphor.

In the *Meno*, Socrates draws figures in the sand to demonstrate that learning is recollection. Yet in a later work, the *Republic*, Plato launches his famous critique of geometry and geometers, only to recant still later when he introduces the pool of science in his parable of the cave of representation. This pool displays naturally formed geometric figures: reflections of the planets—sun (circles) and stars (points). As a reflective device, it is a mediating screen preparing the prisoner for the truth or character of *nous*/sun. This is a necessary mediation because the sun's rays (Ideas) are such that in their brilliance (purity) they will blind the hapless cave dweller. This metaphor of the pool as a once-removed device installs, thereafter, the idea of a connection between mediation and reflection, both of which, in turn, become contrasted with an out-of-sight truth or absent referent.

Proclus' elimination of this hurdle is an inspired move, eloquent in its simplicity. He installs a screen between purity (*nous*) and gross impurity (sensible appearances). This is inspired, for he takes the Euclidean idea of a two-dimensional plane upon which geometry carries out its functions, and marries it with the reflective or mediate character of Plato's pool of science. The result is a mental screen that reflects in part the upper reaches of intellection at the same time as it shares, also in part, certain features of the lower world of appearances.

While this solution is simplicity itself, the full character of this ancient mental screen is far from simple. This is because, notwithstanding Proclus' resolution of Plato's ambivalence, there remains within the above geometric horizon a basic tension between the pure (which can be thought but not imagined) and the impure (which can be imagined but is not properly thought). Laying the ground for his discussion on the geometric imaginary, Proclus speaks of two features that subsequently are never absent from the theoretical structure of geometry—namely, purity and impurity, or in contemporary terms, regularity and irregularity. To understand the role this binary plays in Proclus' screen of the imaginary, we will need to note his description of the most basic element of Euclidean geometry: the point.

THE POINT AND ITS ORIGIN

Proclus begins by broadly defining the geometric point as an essentiality. It is neither an empirical thing nor a pure form in the Platonic sense. As an object of mathematical inquiry, it inhabits the intermediate world of the imagination as already mentioned, that is, a world between the simple, indivisible, and partless reality of *nous* and the composite, divisible and differentiated world of the senses. Proclus' division of reality modifies the Platonic division between the higher world of ideal forms (*nous*) and the lower world of sensible phenomena with the introduction of an intermediate world of the mathematical imagination. Because the point in the imagination lacks actual matter and, therefore, is not susceptible to change as are the objects of the empirical world, it therefore reflects the higher world of the mind. But since it is the final residue of any complete division of geometric figures capable of being carried out within the imagination, it is therefore, like the divisible empirical world, open to perception.

Proclus' notion of the origin of the image of the geometric point in the imagination is equally eloquent in its simplicity. Since a perfect point is obviously never seen in the world of the senses, it must come from somewhere else. Ergo, the perfect point is ideal. Hence, the way the imagination generates an image of the point is by allowing the ideal point to be "projected"[3] down from *nous* onto the "geometric space" of the imagination.[4] This projection is triggered by the mind recollecting or being reminded, as a consequence of experiencing in empirical space, such objects as may be perceived as points. These empirical points are nevertheless impure, Proclus says, since they obviously comprise visible parts. Thus, the perception of impure empirical points calls up the purity of the ideal point, which in turn results in the projection of an image of this ideal point onto the geometric space of the imagination. As *nous* has no time or space, it follows that there is no geometry of the *nous* or of the ideal point.

The significance of this is that geometry is understood as a practice or method bridging or facilitating a movement between ideal and sensible realities. There is no implication in his conception of hierarchical realities that geometry itself is a noumenal form. Rather, it is simply the closest or most approximate form (exemplar) reflecting in a more or less imperfect way the perfection or purity which is characteristic of *nous*. What geometry reflects or brings to our comprehension is not the geometry of *nous*, since there is none, but the perfection, order, and unchangeability of this higher reality. Moreover, since geometric images of points in the imagination are thought to remain more or less the same from generation to generation, it follows that their source must be the world of unchanging principles. They reflect a feature of the unchanging or nonhistorical reality constituting the very structure of the Proclian-Platonic *nous*.

To capture this sense of nonhistoricality or givenness, the point must then be thought as an infinitesimal puncture or dot in which time and space do not exist—a presence in time and space having no time or space. It is a position surrounded by time and space but not partaking of time and space. But, since we cannot actually imagine such a puncture or dot without ever so slightly giving it at least some space, how then are we to think this Proclian point of *nous* without some image? To put this another way: to retain the character of this Proclian divide of higher and lower realities, this *thinking* of the ideal (nongeometric and nonhistorical) point will need to be essentially differentiated from its lower impure images, yet in some manner be capable of being triggered, brought to mind, by exposure to the empirical realm. It must also be capable of being projected onto the imagination in an approximation of the original purity of itself as an ideal point. Can Proclus' ideal point ever be *thought* entirely without the invocation of some image? If not, what image emerges that is not caught in the web of descending approximations that Proclus counterposes to the purity of ideality?

Consider, for a moment, the etymological debt of the term "point." The everyday meaning is: that which has position but no magnitude or parts. Its Latin origin is the term *punctum*: a small dot or puncture, or an instant in time. *Punctum* is the particle neuter of *pungo*, which means to thrust or to strike. One solution to the dilemma of keeping the ideal point pure might be to grant this image of a formative striking the status of an ideality itself. As an ideality, it would by definition be free of being an approximation, thus removing the possibility of an impurity entering our higher thoughts. However, as striking is associated by definition with change in space and time, and as these spatial and temporal features apply only to the empirical world and not to the world of static principles, then to grant striking (mediated change) the status of an ideality would be a contradiction in terms. It would be suggesting that idealities change over time. Can the image of striking be idealized without interfering with his conception that idealities are unchanging and undifferentiated?

What the "point" etymology indicates is that there is an intimate relationship between a mediated act and the immediacy of the ideal. The two are necessarily interdependent and cannot in any absolute way be separated or unproblematically reduced to each other. Indeed, the image (in this particular case, striking) is always intricately interlinked with any *thinking of geometry*. Thus, the problem for Proclus is not so much this relationship as its sequence. For Proclus, in order for the ideal point to be thought of as a pure given and as the originary source of the geometric point in the imagination, then the latter point can only make its appearance *after* this purity. This etymological reversal allows him to install geometry in a maieutic relation to an ideal, pure, unchanging truth. Indeed, the whole purpose of Proclus' commentary on Euclid is to displace what might be called the surveyor's experience of geometry in favor of axiomatica, or first principles, which are presented as being of a higher ontological value than the world of appearances.

However, this relocation is contaminated. Since the geometric point of Proclus' imagination enjoys a certain impurity and is offered in his schema as the final residue of any geometric figure (graphic or imaginative), then rigorous demonstrations of geometry become grounded in a certain impurity—that of the impure geometric imagination. The point is that any privileging of immediacy comes at a certain price. Geometry at the level of empirical application (and its appearance in the imagination) must be constituted as impure in relation to its original generative force. Proclus seeks to reduce this tension through the intermediary of the partly pure and partly impure imagination. The long history of axiomatic geometry, and its philosophical construction as the royal road to truth, thus begins with a certain disavowal of history or the relegating of mediacy to a subordinate position. This relegation permits the highest privilege being afforded to absolute pure or self-evident (nonhistorical or given) truths.

Moreover, at the very outset of the theorization of the geometric imaginary, a geometric process contaminates both sides of the reality divide that geometry constructs. That is, Proclus builds his edifice by employing a *gradient* marking out in degrees of impurity the realm comprising geometry's horizon of measurability (the geometric imaginary) and the realm comprising its horizon of immeasurability (the lower world of the sensible).

GENERATION

Proclus further pictures the imagination as a screen whereupon *nous* writes or projects images. This is a plane of inscriptions and of reflections, a writing tablet and a mirror. To gaze with Socrates' *eye of the Soul* into this mirror reveals to the understanding a picture of itself. Apart from it being written on by *nous*, this screen also writes itself. Since this dual writing gives us a further insight into the complicated structure of the geometric imagination, it is useful to cite the relevant passages in full:

> And thus we must think of the plane as projected and lying before our eyes and the understanding as writing everything upon it, and the imagination becoming something like a plane mirror to which the ideas send down impressions of themselves.[5]

Furthermore,

> If, however, mathematical forms do not exist by abstraction from material things or by the assembling of the common characters in particulars, nor in any way later-born and derivative from sense objects, of necessity the soul must obtain them either from herself or from *nous*, or from both herself and that higher intelligence. Now if she gets them from herself alone, how can they be images from intelligible forms [Ideas]? And how can they fail to receive some increment of being from the higher realities, occupying as they do a middle position between indivisible and divisible nature? And how can the forms in *nous* maintain their primacy as the first patterns of all things (if mathematical concepts are not derivative from them)? Yet if they come from *nous* alone, how can the inherent activity and self-moving character of soul be preserved when she receives her ideas from elsewhere, like a thing moved by outside forces? And how will she differ from matter, which is only potentially all things and generates none of the embodied forms? There is left only the conclusion that soul draws her concepts both from herself and from *nous*, that she is *herself the company of the forms*, which receive their constitution from the intelligible patterns but enter spontaneously upon the stage of being. *The soul therefore was never a writing-tablet bare of inscriptions; she is a tablet that has always been inscribed and is always writing itself and being written on by nous*. For soul is also *nous*, unfolding herself by virtue of *nous* that presides over her, and having become its likeness and external replica. Consequently if *nous* is everything after the fashion of intellect, so is soul everything after the fashion of soul; if *nous* is exemplar, soul is copy; if *nous* is everything in concentration, soul is everything discursively.[6]

Moreover, the soul is constructed "out of all the mathematical forms."[7] Finally, mathematical ideas are self-generating, and their activity in the soul is to bring to the light, more and more, the truth (the partless Ideas) of *nous*.[8]

All told, there are four basic features to Proclus' schema:

1. *Nous* is the realm of idealities (Ideas) that are thought to be primary and unchanging. Since they are the first patterns of all things they hold within themselves the potentiality of all things. An example of this is the ideal point in the condition of possibility for all extension (imaginative or physical). Or in reverse: all extension is reducible to the ideal point. This potentiality of *nous* can on occasion be triggered, by the world of the sensible, into projecting images (impressions) of the ideal point onto the screen of the imagination. In the latter sense, *nous* can be seen, in part, to be a dormant fundament of all meaning thought at the level of the understanding *quâ* Soul.
2. The Soul or understanding has a conceptual and image-making capacity (self-generating). The Soul is also capable of being affected (projected upon) by *nous*. An example of this image-making capacity would be that, from the initial projection by *nous* of an image or impression of the ideal point, the imagination is able to generate (think or image) extension. In this dual manner, the field of the imaginary is always already a mirror (a copy, or written upon by *nous*) and always already self-generative (able to write itself) and discursive (able to write toward the truth: approximations).
3. The forms inhabiting this screen are *not* abstractions from objects of the sensible world. Nor are they an assemblage of common features that objects may share with each other. Moreover, geometric forms do not come after, are not subsequent to, the senses. Rather, they always stand over the sensible, just as *nous* always stands over the Soul/imagination. From this privilege is raised the fourth and last feature, that of origin or source.
4. Proclus leaves in place the differences of these two realms (*nous* and the sensible) without appearing at first to be granting any characteristic of the sensible to the geometric imaginary other than the previously mentioned divisibility-impurity of the geometric point. He speaks of two senses of production, one static, the other dynamic. On the one hand, we have reflection, which he ties to the question of source. The geometrical point is produced, reflected on the screen of the imagination by a projection of the ideal point from *nous*. In being a reflection, the imagination's geometric point approximates the unchanging character of the ideal point. It is in this sense of being merely a reflection of an ideality that this production may be spoken about as static. On the other hand, the inherent activity and self-moving character of the imaginary—its discursive, exemplary, or truth-seeking function—is a dynamic production making possible the actualization of the potential for extension contained in *nous*.

The inherent activity of the imaginary is therefore toward generating figures that approximate the truth of *nous*. They have as their starting point the geometric point, which in turn reflects the ideal point containing the potential for all extension. In this way, the imaginary's activity of generating extension (geometric figures) reflects the potential for extension contained in *nous*.

All this seems very neat and logically consistent, as indeed it is. However, what Proclus does not allude to in any direct manner, is that the inherent generative capacity of his screen is a mathematization of sorts of the generative character of the world of the sensible. While he informs us that the world of matter cannot generate embodied forms (geometry's figures), and that the forms of the imagination come about impulsively (enter spontaneously upon the stage of being)—in the latter sense, we see the beginnings of the role of intuition in geometry—there remains the fact that the imagination is (partially) dynamically productive. If we allow that the world of the sensible is dynamically productive as well, then how is generation in the imaginary not also a mathematical reflection of the disordered generation evident in the realm of the sensible? The answer within Proclus' logic is that the generative ability of the geometric imaginary cannot be thought solely at the level of process. This possibility is ruled out when he says that the Soul is in the company of its forms, and when he says that the Soul is never a bare writing tablet but writes and is written upon (by *nous*). And it is also ruled out when he says that generation is spontaneous.

The upshot of all of this is that, because the imagination produces a certain degree of impurity (divisibility-differentiation), and because it is always in the company of its forms, reflecting more or less the purity of *nous*, then the geometric imaginary as a whole emerges as an approximate field whereupon approximations (impure images) wend their way or point toward a higher, ultimately unattainable truth. Hence, the generative activity of the imaginary is discursive. Conversely, the generative activity of the sensible is not a generation toward truth. It does not generate its own geometry in any systematized way that is endlessly recoverable and applicable. (We will see at the very end of this study a challenge to this ancient idea.) All of this is capped off by having the imagination capable of spontaneous generation, thus removing any direct link with any outside force.

Again, the above is logically consistent and is all very well as far as it goes. There is, however, a seepage in this system that cannot be gotten around solely by making the imagination coextensive with its generated images, reflective of *nous*, and spontaneously generative. We still have to account for the impurity of the geometric imaginary. Let us backtrack for a moment.

IMPURE WRITING

At the end of the above citation, Proclus tells us that the Soul is *nous*. What he means by this is not that the Soul is *nous* in the sense of being completely identical but rather that (in the form of the imagination) it is a reflection, a replica that more or less participates in *nous*. What, then, does it reflect? It reflects more or less the static order and potentiality of *nous*. First, the static order is reflected in the fact that the geometric figures generated can be consistently (spontaneously) repeated in the mind of the geometer. Geometry, of course, is foremost an ordering system grounded in repeatability (the return of the same).

Second, the potentiality for all possible extension characteristics of *nous* is reflected in the actualizing of this potentiality. It is reflected in the self-generation toward the extension character of the geometric imaginary. In this sense, the imagination is also dynamic. Third, the Soul differs from matter because matter is only potentially all things and generates none of the embodied forms. The latter point is a little more difficult than it initially appears, for it functions by default to posit matter (the world of the sensible) as always generative of grossly impure forms (nongeometrical shapes).

It goes something like this: *nous* is potentially all extension. In addition, matter is potentially all things because all sensible things coming into existence must have matter. Equally, Soul reflects *nous*' potentiality by being generative. How, then, does Soul avoid being like matter? Next, matter cannot generate the embodied forms. It can only generate physically impure forms (en-mattered objects). Equally, Soul cannot generate *nous'* Ideas. It can only reflect them impurely in relation to the purity of *nous*. How, then, does Soul avoid being like matter? We are already familiar with Proclus' answers in the above citation.

But, in these answers, Proclus cannot hold the edges of his own schema together. For the geometric imaginary cannot, by definition, get its impurity from *nous'* writing on it, since *nous* is cast as solely pure. Hence, the impurity must come from the second writing. It must come from the imaginary writing itself. However, this inherent and spontaneous capacity to generate impurity is also a characteristic of the world of the sensible. So, although the task of the geometric imaginary is to reflect *nous*, nevertheless, Proclus cannot stop the flow from *nous* to the imagination from continuing down into the world of the sensible. He cannot stop the geometric imaginary's spontaneous self-generation of a more refined impurity from ultimately being perceived as the higher truth of the spontaneous self-generation of the gross impurity of the lower world of the sensible.

Moreover, if we recall his closing comments about the discursive character of Soul, then we would have to allow this unwanted flow—the flow to the sensible—from forcing onto the sensible's naturally generated forms the status of being, at best (gross) approximations of the more pure forms inhabiting the geometric imaginary. (This inevitable relationship between the sensible and the imaginary will continue to play a part in later theorizations of the imaginary.) The lower realm of the sensible therefore emerges as a natural discursiveness, only in an unordered, gross, or more impure manner in relation to the geometric imaginary. Just as Soul keeps the company of its forms, so, too, does the sensible keep the company of its forms. Therefore, the processes of generation in the world of the sensible cannot be separated from the forms this self-same world generates, any more than the imaginary can be thought separate from its contents or, for that matter, any more than *nous* can be separated from its potentiality for extension.

CREATIVITY

If we allow that generation and creativity have anything in common, then for Proclus to have schematized generation is at the same time to advance through the back door a theory of creativity. This is a theory grounded in a gradient, and marked off in degrees of impurity. First, creativity at the level of *nous* is in the form of potentiality, which is a kind of pure or nongenerative generation. Second, at the level of the imaginary, creativity is spontaneously generative but is directed toward the actualizing of the before-mentioned potentiality in a set of impure figures of extension: geometricals. Third, creativity at the level of the sensible is the gross or abundant materiality imbued with the "potentiality of all things" but which "generates none of the embodied forms." Yet it clearly is, by default, also thought to generate impurity, which in turn is more impure than the impurity of the geometric imaginary.

Since the geometric imaginary is positioned as exemplar, it follows that its impurity cannot escape being also the higher truth of the generation of gross impurity at the level of the sensible (although Proclus has excluded the geometricals from being abstractions from the sensible). If we play a little joke on the logic of Proclus, if we allow the flow to go both ways, then the sensible becomes the exemplar or preparatory way in which to glimpse the greater purity of the geometric imagination. This is a process somewhat reminiscent of the purpose underpinning Socrates' drawings in the sand. The irregularity of the objects in the world serves to remind the viewer of the greater perfection generated in the imaginary.

Indeed, although somewhat in its infancy in the above schema, the possibility for the mathematization of creativity does not entirely escape Proclus. This is seen in the following citation, where he describes the parental or formative role that the ideal point of *nous* plays in the generation of the extension of all figures:

> the point,...although its being is determined by the Limit, it secretly contains the potentiality of the Unlimited, by virtue of which it generates all intervals; and the procession of all the intervals does not exhaust its infinite capacity. Being a Limit, [it] preserves its character when things participate in it. But since it also secretly possesses the nature of the Unlimited and strives to be everywhere in the things that it bounds, it is present in them an infinite number of times; and since in the intelligible world the Unlimited is a power generative of extended bodies, so it is potentially such in the things that share it. For among the higher realities—the intelligibles—the Unlimited is the first creative cause and generative power of all things, but in enmattered forms it is imperfect and only potentially everything. All limits exist preeminently in *nous*,...but partless and without differentiation of kind, so that they all subsist covertly and indivisibly in a single form under the idea of the point.[9]

It should not be forgotten that this gradient also generates a certain sterility. For under the mantle of the privilege already placed on order, generation goes from the turbulent creativity manifest at the level of the sensible, first toward the more or less ordered realm of the imaginary, and then on to the static environs of *nous*, manifesting itself there as only potentiality. However, inasmuch as Proclus' schema resembles Plato's theory of representation, it is important to remember that Proclus (like Plato) is not claiming that the world of appearances is an illusion. Ideal reality is not the opposite of the reality of appearances. Appearances are merely a deficient reality because they hide in their abundance or confusion the eternal principles or laws governing nature. More precisely, they hide the eternal and static truths of *humanity's* Being. (Also, bear in mind that both Proclus and Plato are not claiming that nature in itself is deficient, but rather that only sensibility *quâ* representations of nature are deficient.) Only by thinking beyond the limits imposed by the senses can the truth to nature, or *humanity's* true reality, be grasped or understood. Ideality is, therefore, the genuine, eternal, unchanging state of true reality. Appearances, by contrast, are a deficient, always changing, less genuine or lower state of reality.

To follow this a little further before returning to the geometric imaginary: What happens when, say, a visual practice such as art is informed or motivated by this idea? The first thing that occurs is a separation between reality and representation. As representation in this schema initially belongs to the concrete world of appearances—this is one of the main themes of Plato's cave story—it follows that representation is separate from true or ideal reality. Given this separation, the task automatically falls upon the shoulders of representation to represent this true reality beyond mere appearances, beyond the limits of sensibility. For Plato and Proclus, the exemplary disciplines for this task are mathematics and geometry, which, while still representations themselves, are nonetheless the purest forms available.

Moreover, there is a further burden or responsibility placed on representation. Because representation is contrasted with the true reality of ideal nature, a reality constituting the higher reaches of intellection, the burden or task for representation, at the lower levels of the imagination and perception, is to reflect adequately this higher or more genuine reality. However, as representations in the form of images will also be more or less deficient, precisely because they are images and therefore limited (have sensible edges, so to say), the task is to find, discover, or create representations less deficient than their predecessors. These are representations that will more accurately reflect *humanity's* (later in the history of geometry, nature's) eternal and unchanging character of Being, which lies beyond the horizon of the sensible.

Another characteristic of this system is the opposition between geometric forms (later known under the rubric of abstract forms) and so-called organic, natural, irregular, or nongeometric forms. For instance, if appearances are composed of organic forms, it follows that the closer to organic or natural forms a representation is, the more likely that it is deficient in reflecting universal laws. By contrast, the more unlike a natural form a representation is, the better it truthfully reflects universal principles. The problem in this system for visual practices such as the plastic arts, is that the closer their representations get to geometric forms, the less likely they will be considered as being *Art*, or the work of any specific individual. Alternatively, the closer they get to organic or natural forms, the less likely they will be considered the reflection of any higher truth of Being. In short, this schema's program of representation is related to the accuracy with which a representation is considered to reflect the truth of some universal principle or ideal. Visual practice can only, at best, strive to produce better and better representations of this ideal or higher reality. It can never, by definition, ever fully complete its task, simply because of the assumed gap between itself and true reality.

However, one way that art can overcome losing its identity by looking too much like geometric forms is to construct a system of representation based on mathematical ratio or proportion. In this way, nongeometric, so-called natural, forms may be used. The emphasis in the representation is therefore concentrated instead on achieving the correct or truthful ratio, proportion, or relationship. Some of the art of the Renaissance has this feature: being a precisely laid-down mathematical system of ratios and proportions at the same time as it uses natural-looking shapes in its compositions.

Finally, as the pressure is on representation to come up with better and better models, and as it (and, to a lesser degree, geometry and mathematics) can never, because of its once removed separation from reality, ever hope to reflect absolutely and finally the true reality, then it follows that representation is always under scrutiny to demonstrate its novelty and adequacy. Change, therefore, is coded as the change from one form of representation to another, better, and newer form of representation.

This state of affairs has spilled over into modern scientific experimentation, technologies, and scientific theories. The ongoing task or challenge is to find better and better models with which to represent and interrogate some supposedly aboriginal reality, or absent referent. Hence, limitations in explanatory power are seen to be a corollary to the perceived limitations of the model employed. The advantage of this system is that, since models by definition are thought to be inherently limited, they also become self-generative: their inherent limits constitute the legitimation and the condition of possibility for the production of newer and newer models. Their inherent inadequacies drive their constant manufacture. That they make their own future, might be another way of saying something similar.

CODA

Proclus' formative theorization of the geometric imaginary has a contaminative structure both at the level of its production or generative capacities, and in relation to the upper and lower realities it interposes, or stands between and separates. Theorized initially as more impure than *nous* while being more pure than the realm of the sensible, its impurity and the possibility of its repeatability (its spontaneously self-generative character) becomes the exemplary way to glimpse the static character and inalienable potentiality of *nous*. Thus, *nous*, in all its purity, staticness, and potentiality for extension, can only be got at *through* the portal of impurity and the always attending, spontaneously self-generative acts of the geometric imaginary. In this sense, impurity (geometry) contaminates or mediates how purity (*nous*-truth) is to be thought in this early moment in Western geometry's history.

The Platonic exemplar (thinking geometrically)—the royal road to truth for 2000 years—thus sets out on its long journey from within a theoretical structure whereby this self-same exemplar stains idealities. The road to the ideal is always mediated by a geometry initially cast as less pure than that to which it leads. By recalling impurity that is also a feature of the world of the sensible, the often remarked irreal world of Greek geometry may be less irreal than previously thought. The exemplary way (geometry) not only stains ideality with its own impurity but is itself stained with the pathology of the lower world of the sensible: Proclus' imaginary is divisible in character.

Moreover, in this ancient theorization, it is only in contrast to the self-generative character of the geometric imaginary and its always attending impurity that the world of the sensible may be thought of as always generative of the grossly or abundantly impure objects of nature. In this sense of the imaginary's imperfect purification of generation, impurity (geometry) contaminates or mediates how gross impurity (nature or sensible generation) is to be thought. Also, the geometric imaginary, by repeatedly reflecting in an ordered fashion the potentiality for extension of *nous*, and by repeatedly reflecting the process of generation (coextensive with its products), emerges as a theory of creativity at all its levels: potentiality, austerity, and grossness. This is a theorization of generation-creativity grounded in a certain pathology or relative impurity at one end and grounded at the other end in purity.

NOTES

1. Regarding the sources that Proclus drew upon for his commentary on Euclid, see D.O. Meara, "Proclus' First Prologue to Euclid: The Problem of Its Major Source," in *Gonimos: Neoplatonic and Byzantine Studies*, eds. J. Duffy and J. Peradotta (New York: Arethusa, 1988), pp. 49-59.
2. Proclus, *Commentary on the First Book of Euclid's Elements*, trans. G. Morrow (Princeton, NJ: Princeton University Press, 1970), p. xxxv.
3. Proclus, *Commentary*, p. 11.
4. Proclus, *Commentary*, pp. 40-45.
5. Proclus, *Commentary*, pp. 97-98.
6. Proclus, *Commentary*, pp. 13-14. Morrow's footnote 27 is inserted in parentheses; my italics and inserts in brackets.
7. Proclus, *Commentary*, p. 14.
8. Proclus, *Commentary*, pp. 14-15
9. Proclus, *Commentary*, pp. 72-75.

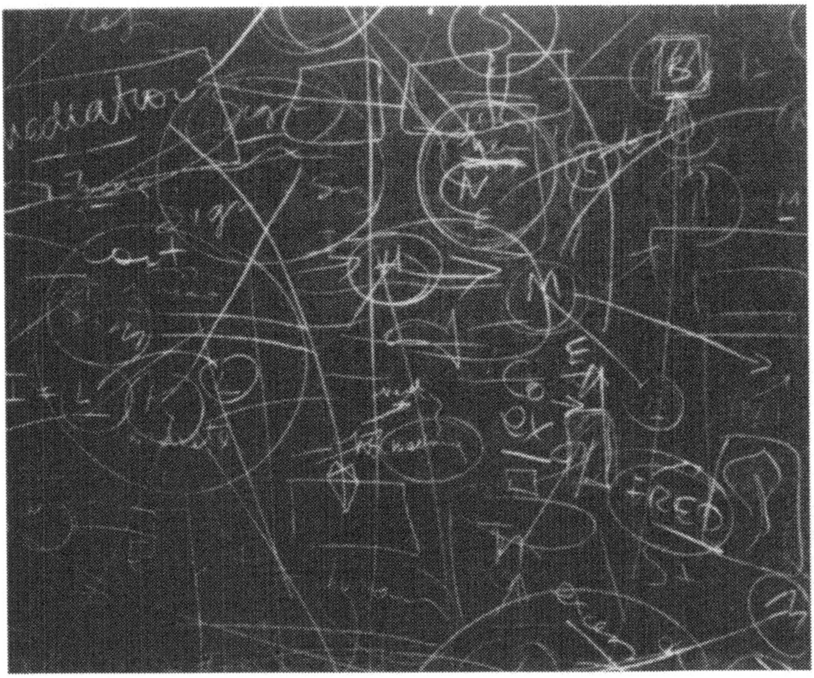

Illustration 5. The cacophony of a blackboard geometry created by Noël Gray and Friedrich Schmidt in September 1994 to outline a general theory of meaning.

Chapter 3

MICHEL SERRES ON CACOPHONY IN GEOMETRY

PATHOLOGY

Proclus does not devote a great deal of space to the exploration of the precise graphic, or physical appearance, of the geometric diagram or figure. Nor should we expect him to, since his interest is expressly in discussing the origin and processes of the formation of geometric images in the imagination and in explicating Euclidean geometry as a whole. It is enough for him that the reader be assumed to be familiar with the basic appearance of whatever geometric figure is under discussion. While he clearly defines the components or elements of the imagination's geometric figures within a gradient of impurity having at its lowest degree the empirical, it is not explicitly the precise relation of the empirical figure to the imaginary figure that he takes as his concern. For this, we need to turn to a contemporary theorist of geometry.

The French philosopher Michel Serres introduces to us an ancient Greek term, "cacography." Cacography is to be understood in the sense of *bad-writing*. He uses the term in reference to a geometer deliberately overlooking the irregularity in the graphic rendering of geometric figures. The geometer is obliged to do this so as to attain the ideality of, say, a circle, gaining the idea or image in the mind of a perfect circle in spite of the countless irregular circles constituting its empirical instances. Pure figures are made manifest in the mind by thinking or seeing beyond the materiality of the drawn figure. In this initial fashion, and contrary to Proclus' sense of its inauguration, purity now comes after impurity, not before.

To illustrate this, Serres begins with the conditions for the possibility of communication.[1] He says that it is important to recognize the significance of the physical appearance of the vehicles charged with conveying the communication in question. For instance, at the level of the reception of a written or graphic communication, it is essential that both parties possess the same key for the coding and decoding of the intended message. This possession is the condition for the correct reception of the meaning. However, at the source of the communication, there is another condition: the drawing should be executed as well as is possible.

Serres then differentiates the drawing into essential and inessential elements. The essential elements he perceives as carrying the syntaxical and morphological rules that give the drawing its meaning. The inessential elements are what he regards as accidentals:

> those without meaning whose presence depends on the ability, the clumsiness, the education, the passion,...the illness of he who writes: waverings in the graphic forms, failures in the drawing, spelling errors, and so on.[2]

The task of these rules is to protect or preserve the meaningful elements of the drawing against the accidents of rendering. The former are the caligramical aspect of form (visual grammar); the latter, the cacographical or pathological. This pathology of communication is not only restricted to the graphic. It makes its appearance in all areas of communication: "stammerings, mispronunciations, regional accents, dysphonias, and cacophonies," and in respect to technical means as:

> [b]ackground noise, jamming, static cut-offs, hysteresis, various interruptions. [Moreover,] to write badly is to plunge the graphic message into this noise which interferes with reading, which transforms the reader into an epigraphist. In other words, simply to write is to risk jumbling a form. [Hence,] communication is a sort of game played by two interlocutors considered as united against the phenomena of interference and confusion.... *To hold a dialogue is to suppose a third man and seek to exclude him*; a successful communication is the exclusion of the third man.[3]

Basically, in any communication there are two interlocutors who are in a symbiotic relationship, the purpose of which is to attempt to expel a third. To establish meaning or understanding, it is necessary to purge the drawing of noise, to overlook its cacographic element. The production of meaning is thus inevitably tied to a struggle against interference, noise, and irregular rendering. Serres offers the example of the geometric diagram drawn on a blackboard. A blackboard drawing is understood by mathematicians as representative of the same symbol, even though every drawing is different. The mathematician is able to reason, to call up the parent symbol or referent not by "using the concrete graph drawn on the blackboard, here and now, but rather, as Tarski says, by using the class of objects having the same form."[4] The graph or diagram merely evokes the symbol or referent.

This symbol or:

> abstract being is recognized by the homeomorphism, if I dare say so, of the graphs. The recognition of this being presupposes that we distinguish the form of what I have already called cacography.[5]

And finally:

> no graph resembles any other and...if we wonder what are the respective portions of form (containing meaning) and of cacography in writing, we must flatly admit that noise prevails...that it prevails exhaustively...[Thus,] *it is one and the same act to recognize an abstract being through the occurrences of its concrete, standardized form* and to come to an agreement about about this recognition. In other words, *the act of eliminating cacography, the attempt to eliminate noise, is at the same time the condition of the apprehension of the abstract form* and the condition of the success of communication.[6]

Serres establishes a necessary relational character of noise and symbol, a relation of necessity between the irregularity or cacographical form and the regularity of the ideal form. It therefore follows, as indeed he says, that the world of mathematics is the realm where all but the minimum noise remains. It is the realm where the "third man" is all but totally excluded. Geometry is the realm of the minimum necessity of relation—where the necessary relationship forged between irregularity and regularity, noise and meaning, cacography and communication, finds its smallest expression, is held at its lowest level. This realm of the minimum relation is clearly reminiscent of the Platonic-Proclian idea of geometry as the exemplary procedure for grasping the reality of idealities. Geometry, purged *almost* completely of noise approaches, comes closest to, approximates, the higher reality of the ideal form. Serres duly acknowledges this debt by saying that the birth of "scientific mathematics" is the moment that made the "recognition of the abstract form and the problem of the success of the dialogue coexist."[7]

Indeed, alluding to the *Meno*, Serres says:

> When I draw a square and a diagonal in the sand, I do not in any way want to speak of this wavering, irregular, and inexact graph; I evoke by it the ideal form of the diagonal and of the square. I eliminate the empirical, I dematerialize reasoning...I create the possibility of a science in the *Universal for us*...and therefore *the first effort to make communication in a dialogue successful is isomorphic to the effort to render a form independent of its empirical realizations*....In this sense, the reasoning of modern logicians concerning the symbol is analogous to the Platonic discussion of the geometric form drawn in the sand: one must eliminate cacography, the wavering outline, the accident of the mark, the failure of a gesture, the set of conditions that ensure that no graph is strictly of the same form as any other.[8]

Serres is specifically granting a staticness to meaning at the level of the ideal form. Hence, at the level of the empirical, differentiation needs to be excluded in order to see the sameness hidden or veiled by the abundant vagaries of graphic renderings. Moreover, the object open to perception in contrast to the ideal form needs to be "indefinitely discernible."[9] Or else, there would have to be a different word ascribed for every perceptual instance of an object, and for every perceptual instance of an object, a different word that took account of the subjective state of the observer along with designating the identity of the observer, and so forth. It follows that, at the "extreme limits of empiricism, meaning is totally plunged into noise," hence, "the first third man to exclude is the empiricist along with his empirical domain."[10]

RADICAL FORGETTING

To follow Serres is to understand that regularity, purity, or ideal staticness is intricately interwoven and interlinked with irregularity. Meaning at the level of its universality is precisely and irrevocably dependent upon the vagaries of the empirical. Yet, at the same time, these self-same vagaries need to be eliminated in order for the ideal referent to reveal itself. How, then, are we to think a relationship of necessity that also demands that the empirical half of the relation be eliminated, or at best kept independent from the ideal half? How, then, is the formation of this Cheshire cat to be accomplished?[11] Serres solves this difficult problem (how to attain the ideal by eliminating yet retaining the empirical) by speaking of a kind of radical forgetting at the level of perception:

> And this demon [the empiricist] is the strongest demon, since one has only to open one's eyes and ears to see that he is master of the world....(And, as has often been seen in any discussion between an empiricist and a rationalist—Locke and Leibniz, for example—*empiricism would always be correct if mathematics did not exist.* Empiricism is the *true* philosophy as soon as mathematics is bracketed. Before the latter imposes itself and in order that it may be so, one must *want* not to listen to Protagoras and Callicles—because they are right. But the more they are right, the less we can hear them: they end up only making noise)....Consequently, in order for dialogue to be possible, one must close one's eyes and cover one's ears to the song and the beauty of the sirens. In a single blow, we eliminate hearing and noise, vision and failed drawing: in a single blow, we conceive the form and we understand each other. And therefore, once again, the Greek miracle, that of mathematics, must be born at the same time....To isolate an ideal form is to render it independent of the empirical domain and of noise. Noise is the empirical portion of the message just as the empirical domain is the noise of form.[12]

The ancient legacy of the power of scientific geometry is again apparent: the promise to bring forth the ideal world against the cacophony of the empirical. The Greek miracle releases us from what is presumably the prescientific, pre-Greek world of chaos, of disorder. It releases us from a dependence upon the empirical as true philosophy. Or, in the modern idiom, empiricism's appeal to itself cannot issue up the truth of itself. Or, as Serres himself says somewhere else, it releases us from the pre-Greek association of truth with perception. However, Serres is not a zealous idealist. Quite the contrary. To understand how he avoids this trap, we need only recall that he needs to retain the empirical, since it is in his schema a necessary element in communication, and that he needs to eliminate the empirical to increase the degree of communication possible. The problem he faces is how to see only the cat's grin without killing the cat.

The simplicity of his solution is impressive: forgetting the noise, overlooking the cacographic character of the graphic, does not destroy its materiality or its necessity within his structure of communication. Rather, it simply erases it from the consciousness. Close one's eyes and it more or less goes away, yet is still there. This is the essence of his solution: a deliberate or radical forgetting of what is already known or seen. Attaining the ideality, getting at the universal meaning, is based on a demand to forget, a demand to overlook.

Or, to once again note a Platonic debt: in the *Meno*, the slave discovers his forgotten knowledge by further forgetting. He is called upon to overlook the irregular character of sand, the desert screen; to overlook the irregularity of what is drawn in the sand. The prescription for attainment of insight is made clear: the slave remembers by forgetting. The sand drawing releases the Idea by erasing, in a manner of speaking, itself. For Serres, drawing the circle in the mind is at the same time to erase the circle in sight. Conversely, to concentrate on the irregularity of the drawing is at the same time to erase the pure circle in the mind: recall his humorous critique of empiricists as ultimately noise makers.

NON-MEANING

Having begun by having purity follow impurity, Serres thus takes a further step to bring his schema in line with the ancient privileging of purity. The necessity of the relation of noise to meaning is a demand for the viewer to overlook pathology. "One must *want* not to listen." Noise has no meaning (recall: noise is without meaning whose presence depends "on the ability, the clumsiness, the education, the passion, or the illness of he who writes: waverings in the graphic forms, failures in the drawing, spelling errors"). The net result is that meaning is produced by expelling non-meaning.

Yet, this is not quite all Serres is saying. For noise at the level of the graphic, he adds, is produced by the Subject (although perhaps accidentally). One must eliminate the pathology, the corporeality marking the presence of the Subject. One must eliminate cacography, noise, wavering, the failed gesture. Furthermore, his desire is to elucidate the communicative, the shared, which, at its highest level is an ideality, or pure figure. Yet, the transubjectivity implicit in this demand is dependent upon an erasure of the marks of the Subject, upon the erasure of the pathology of the drawing. Thus, ideality comes at the price of the Subject. It comes at the price of the erasure of the very thing Serres also needs to invoke to make erasure possible. Put another way, the Subject who must "*want* not to listen" is the self-same Subject who enacts the pathological utterance, who enacts the wavering outline.

This bond of necessity grounding the possibility of meaning therefore turns out to be *not* one forged between message and noise, meaning and non-meaning (for how can non-meaning be?). Rather, it is a pact predicated upon a divided Subject. Non-meaning noise, or the pathological trace of the Subject, is as much inscribed with meaning as that to which it is counterposed, namely, the reasoning Subject. Indeed, for Serres to speak of the diagram as pathological is already to ascribe meaning to it. To go one step further: we may now see the full import of Proclus' intervention in the Platonic schema of graded realities, because he brings to our attention that mental images, those in his screen of the geometric imagination, are to be understood as impure, irregular. Proclus' mental geometric figures thus have a pathology; they suffer from a mental or conceptual cacography.

In light of this ancient insight, Serres' thinkers need not close their eyes to see ideality, anymore than as empiricists they can elect to stop thinking. Quite simply, neither are possible in any exclusivity, nor are the former in "seeing" perfect purity. The transcendent image or universal symbol seen in the mind is a relative of what is seen with the eyes. While Serres is emphatic about acknowledging difference homeomorphically, he is less inclined to take the next step and see that the mental modality constitutes the same difference in relation to not being an empirical image. Or, in reverse fashion: the pathology of the graphic figure is nothing more or less than a concretization or materialization of the pathology of the mental image.

Indeed, given Serres' divided Subject, both mental and empirical images must be thought of as cacographically constructed and constituted. Hence, Serres emerges as actually talking about different forms of cacographic utterances, in contradistinction to his desire to elucidate the formation of a noncacographical shared universal realm (contrasted with the empirical cacographic chains defining the local). Put another way: Where does Serres show us that the Subjects who *think* the homeomorphic form are able to free themselves from their cacographic tendencies, from their illness, their passions, their (mental) clumsiness? By what manner do cacographical utterances stop at the level of the empirical, stop from contaminating the upper reaches of Serres' pure reality of the mental? Conversely (and to borrow from his lexicon): to think or imagine is *also* to risk jumbling a thought or mental form.

What Serres' schema thus brings to the fore is not the essential truth to communication (whatever this might finally turn out to be), or the proof of some cross-cultural form or universal symbol. Rather, he brings to our attention a set of particular mechanisms of a certain kind of privileging grounded in a philosophical debt. This is a privileging that grants the truth to space as *regular*. And one that argues for an ideal image standing inside all those irregular images executed at the level of the empirical.

Alternatively, by now understanding cacography as contaminative or coextensively present in the empirical and the ideal, it is possible to say that the irregularity of the visual image contaminates the regular image of the mind. It is, in fact, what gives the image in the mind its appearance of regularity. Likewise, the conception of regularity gives the visual image its appearance of irregularity. They work hand-in-hand to define each other. They cannot ever be absolutely separated from each other, even though traditional geometry seeks to privilege mental *regularity*, in contrast to the *irregularity* of materiality. This, of course, is not to say that they are not different from each other, merely tht they are not absolutely different in this difference.

In short, although the pathological function in Serres' schema is intended to project a privilege onto ideal or mental forms, he also has specific graphic figures sharing something in common (a universal symbol): their "circleness," their distinctive "shape" that is clearly discernible in spite of their pathology. Certainly, this is how I understand his remark concerning the homeomorphism of graphs. However, although he calls the homeomorphic form a symbol or abstract form, it needs to be asked: in what form is this form, this distinctive shape, attainable to the mind if not as an image, and as such, is it not therefore prone to conceptual cacography, prone to mental clumsiness, prone to a certain pathology? Indeed—and to widen the horizon of this study for the moment—any appeal to cognitive processes or patterns of recognition grounded in morphological rules as something that gets around cultural specificity, or that is transcendent to any specific culture, specific language, or individual traits, falters precisely because it can never shed its necessary dependence upon cultural meaning. Which, in this instance, is as irrevocably tied to whatever we call cacography as that which is then opposed to the homeomorphic element of these self-same utterances. To see the *sameness* in any form is already to acknowledge cultural differentiation, rather than to witness some transcultural form.

To argue otherwise would be to say that all cultures are more or less Western. Which, if we are to follow the logic of all the citations discussed so far, would be to deny ultimately the Greek miracle (of discovering the sameness under the veil of abundant appearances). It would deny that which Plato, Proclus, and Serres et al., have mobilized as that which denotes the origin of the Western scientific and technological culture. In another fashion: to argue that homeomorphology reaches out, beyond, or over culture, is to subsequently erase this miracle. It is to forgo the specificity of Greek scientific geometry by arguing for transcultural cognitive processes or patterns of recognition grounded in universal morphological rules. This matter of cultural specificity versus geometry-as-transcultural goes to the heart of science's efforts (past and present) to speak of the experiential. It turns on the question of the edge.

EDGES

Briefly, what an edge is, is a relatively straightforward question when we are only meaning to determine the particular limits of some inanimate object. However, when we are speaking about something as significant and volatile as culture, subjectivity, identity, or any other process with deep epistemological and ontological debts and profound influence on the extent, character, and condition of lived experience, then the innocuous edge becomes a matter of some import, at the level of both its corporeality and its ethereality.

Indeed, whatever can be said to constitute a cultural edge is always going to be a volatile and contested matter, not least because the term "edge" has at a vernacular level a strong material and finished sense about it. It therefore seems slightly inappropriate when used in conjunction with the word culture. Equally, the term "culture" does not of itself exclude the material. Quite obviously, culture has a material face—although the entire question of a culture's edge would be redundant if all there was to culture was an exclusive materiality. And, in speaking of material borders, we do not need to be reminded that to mark such a physical edge has enormous political ramifications, as does not to mark it. With these difficulties in mind, I will try only to give a little edge to the question of geometry and our culture's edge.

For instance, staying within the logic of traditional geometry, if we continue to adopt the long tradition of posing intellection and perception as divisive and uniquely separate ways of engaging reality, in what way, then, may we think or see the edge of something as mercurial as culture? In what way may we think the edge of something that in a sense we presuppose by already speaking of it as "something"? A word that clearly suggests a certain finitude, a certain measure that, in a sense, precedes whatever measure we are yet to conclude will determine the edge of this same "something."

Even putting aside these presuppositions and the specter of an infinite regress, we should still be mindful of the formidable task of questioning a culture's edge. To ask this question is to be already driven to acknowledge that we are deeply embedded in and always inseparable from the very thing we are seeking to determine the edge of. Hence, there is no outside position here that would make sense of speaking about the edge of culture in the way; for instance, we speak about the edge of some object, notwithstanding that objects are equally saturated with culture. Nevertheless, we do in fact speak in the latter way as though we were in some incontestable fashion outside the object. So, let us therefore go along this path, if for no other reason than to seek to glimpse the difference here in relation to giving an edge to a tactile inanimate object in contrast to speaking about giving an edge to culture.

This small set of qualifiers, concerns, or theoretical edges may be summed up by saying that it is always already a cultural question to ask what is the edge of a culture. It is a question that only makes sense to a Being already immersed in, profoundly enfolded in, and profoundly contaminated through and through by culture. To argue otherwise would be to adopt the position of a transcultural Being able to be outside culture, yet one still somehow able to ask this specific cultural question. This would amount to the claim that what can only be a radically unknowable question in all its cultural fullness to *the one outside*, is nonetheless somehow also knowable in a way that still leaves intact that the knower be outside. Hence, whatever could be knowable to any outsider would thus emerge as just the edge of its own so-called outsideness. It certainly would not be otherwise unless our so-called transcultural Being could in some fashion be both inside and outside that which it is radically other to, in order initially to pose itself as transcultural.

Moreover, let us also note that we in our turn as cultural Beings are only correct when we speak of culture as being that which we ourselves understand as the processes we ourselves are embedded in. It is, therefore, not strictly correct to speak of the edge of some *other* culture, for in so speaking we are only really defining the edge of our own. The latter difficulty is an old debate that goes under the terms translation-transformation and cultural commensurability versus cultural incommensurability. The last term connotes a (negative or yet to be found) measure able to determine discreteness. That is, incommensurability contains implicitly the sense that the other culture which is incommensurable to ourselves may be made commensurable *if only* the absent measure can be found that will reconstitute difference to some index or gradient of the same. This is so even though we may sometimes feel that such a commensurating gradient is possible simply because we can observe a Being in another culture carrying out, say, an act of measuring, and hence we are led to suppose that to measure is universal.

However, this idea would only function coherently if we were able to separate event from meaning. Because, while we can conceivably subsume a particular event in another culture under the mantle of our own idea that measuring is a universal, it is something else again to suppose that this observed event is understood as an act of measuring by whoever in the other culture is involved in orchestrating it. Or, for that matter, that it is even taken to be an event in the way we understand such a thing (or even that the orchestrator is thinking of him or herself as an individuated self, and so on). Conversely, to know the event and its meaning as they are possibly knowable to the orchestrator in another culture is also a nonpossibility. It is always unknowable to us in any sense of its original ontic validity (if, in fact, it has one). We can never shed the fact that all events we perceive (regardless of what we take to be their initial generative origin or source) are always tied inexorably to our own meaning and its production. They are inexorably mediated through, contaminated by, our own meaning system.

In short, everything we perceive (regardless of and in spite of its so-called "other" cultural source) will always be mediated, including what we may choose to regard as transmediate, transcultural. Even to perceive an event that we immediately understand as an act indicating the universality of measuring is already to give it meaning. This is so regardless of how immediate the perception of the event may seem. To imagine otherwise is to be entranced by our own valuing system. It is to be entranced, on the one hand, into believing that immediacy is outside mediation and, on the other, into forgetting that perception is always already value-laden. Even to speak of an event as transcultural, transmediate, or immediate, is already to be enmeshed in some theory of mediation (at the same time that one is attempting to displace it by invoking the concept of the transcultural, transmediate, or immediate).

Alternatively, for the universality of a universal to be truly demonstrated, such a demonstration would have to take place outside all and any meaning systems. It would have to transcend language in the fullest sense of this term. It is difficult to imagine how this might be accomplished without ultimately mobilizing some form of representation specific to our own culture. It is, therefore, not a trivial thing to say that the universal is for us always inside meaning, not outside. The universal is not a thing that language merely refers to (some non-mediate referent). Rather, it is a kind of surface trace if you like, of an essentializing and ideation process enfolded in our meaning production. To be sure, universals are necessary for us to think an essence or unchangeability that we might feel characterizes our existence (as it endlessly drifts and darts in a sea of ceaseless change). But this essentiality in its essentialness is entirely different to being transmediate in respect to different cultures. Therefore, cultural transmediation only gains its sense if mediation, meaning, can be suspended. This leads again to the question: In what way can we think outside meaning? In what way would we represent to ourselves the cultural transmediate without using representations intricately enfolded in our peculiar culture?

Given, then, this inescapable difficulty of an inseparable immersion and influence by culture, what are we in this culture asking when we ask What is the edge of culture? Well, one of the things we seem to be doing by even speaking of an edge is to implicitly or explicitly mobilize rigorously or casually some measuring system, some geometry with which to define what will count as an edge, and then to plot this self-same edge. Bearing in mind as we go about this task that geometry is as much a generator of meaning as it is a tool with which to discover meaning, what edges then does geometry generate?

Within the scope of traditional geometry, and disregarding for the moment what I have previously said about the geometric imaginary, the first edge generated is a negative edge. To understand this strange near-contradiction in terms, we need to backtrack a little. Recall that for Proclus and Plato, the highest level of the image is the imagination, which in turn logically leaves the upper reaches of intellection image free. This blind intellection is caused by the fact that *nous* or the upper reaches of reality, the realm of idealities, is an undifferentiated and unchanging whole, a partless realm partaking of no division. What all this means for our concerns is that geometry announces its presence in the imagination, and not in the Platonic upper reaches of intellection. Therefore, if we are to speak coherently about culture as having an edge, we cannot expect to locate this edge in Platonic pure thought; that is, Platonic pure thought partakes of no image, is indivisible, partakes of no geometry. However, since we can still think the idea of culture as having an edge, then in so thinking we must admit that it is an idea that exhausts the image-making capacity of the imagination. Unfortunately, in retaining coherency, this tends to install culture as an ideality, installing culture as unchanging. Alternatively, we can accept that culture's edge is only to be thought at the level of the imagination.

The first move leaves culture with a negative edge, an edge that is just beyond the imaging capacity of the imagination. Which is to say that culture's edge is bounded by what is not only a blind intellection but also now a blind imagination.

However, we should not dismiss too quickly this move toward ideality. For without doubt, there is a strong need to retain some sense of a transcendent aspect within our culture at the level of the experiential, if for no other reason than to avoid a fall into radical empiricism in its most prosaic form. The second move leaves us with the problem of trying to picture in our imaginations an actual image of the edge of culture. In geometric terms, we are confronted with giving a definite shape to culture. It is precisely here that we seem to reach the end of the capacity of the traditional geometric imaginary to come to our aid. Clearly, in continuing to put aside cacographic contamination, none of the plethora of austere shapes available to traditional geometry appear to describe adequately what we have perhaps already decided is at the very least a mercurial, always changing condition.

I use the term "condition" decidedly. If there is a way out of our dilemma, then it is assuredly to adopt the idea that culture, if it is any one thing at all, is more likely to be a field of processes or mercurial conditions, rather than some static object reducible to any of the forms of traditional geometry. Culture, precisely because it comprises qualitative elements, is not something that can easily issue up an ideal (austere geometric) shape the way the quantitative procedures of traditional geometry have sought to do with the multitude of disparate forms perceivable in nature (or inherent in *nous* as undifferentiated). Culture in all its fullness—which is to say, in geometric terms, in all its irregularity or dynamic character—simply escapes the overt static regularity of traditional geometry, even though the geometry of austerity or axiomatica was borne in the very heart, perhaps at the very emergence, of this self-same culture. And having reached this edge, I will leave the question unanswered.

CODA

Insofar as the character of the geometric imaginary has been brought forward at this point, what has emerged is that in Proclus' theory, the horizon of the imaginary is bounded at one end by purity and at the other by impurity. His geometric imaginary reaches out in one direction toward a horizon of measurability and in the other direction toward a horizon of immeasurability. In Serres' case, the structure of his imaginary is almost identical, except that its horizons are bounded with non-meaning or noise at one end and by meaning at the other.

Their conceptions of the separation of their respective horizons, while differing here and there, are nevertheless also basically similar, since the very idea of separation permitted them to privilege one horizon over the other. Equally, however, in both their schemata, this separation becomes difficult to sustain. In Proclus' case, a seepage throughout his divided realities took place once it was noted that his theorization of the geometric imaginary also amounted to a theorization of generation-creativity: first, creation as initial condition (*nous*' projection); second, creation as the actualizing of extension (the imagination's writing); and third, creation as gross production (sensible matter). This presence of a theorization of creativity at all levels was further augmented by his employment of a geometric process in the form of a gradient marked by purity, impurity, and gross impurity. In Serres' case, his schema also leaked when it became apparent that his Subject inhabited both sides of his divide of meaning and non-meaning. This was also apparent when cacography, pathological behavior, could not plausibly be held solely at the level of the material rendering of the graphic image or diagram.

In sum, in both these geometric imaginaries, the relation among separation, privilege, and generation-creativity is at the very heart of traditional geometry's conception of itself as the reflector of purity. However, when passed through the processes of contamination (derivable from within these theorizations themselves), what emerges is this ideal world, reflected in ever-repeatable procedures, that is stained with a certain pathology, one marked out on a gradient. Putting these features together, we may reasonably say that traditional geometry is *a pathognomonic science*. More precisely, essence or purity in traditional geometry is more or less pathological; geometry's measure is inherently irregular.

NOTES

1. Michel Serres, "The Origin of Geometry," in *Hermes: Literature, Science, Philosophy*, eds. Josué Harari and David F. Bell (Baltimore, MD: John Hopkins University Press, 1982), p. 65.
2. Serres, "Origin," p. 66.
3. Serres, "Origin," pp. 66-67.
4. Serres, "Origin," p. 68.
5. Serres, "Origin," p. 68.
6. Serres, "Origin," p. 68. Italics mine.
7. Serres, "Origin," p. 69.
8. Serres, "Origin," p. 69.
9. Serres, "Origin," p. 69.
10. Serres, "Origin," p. 70.
11. "I wish you wouldn't keep appearing and vanishing so suddenly: you make one quite giddy!" "All right," said the Cat; and this time it vanished quite slowly, beginning with the end of the tail, and ending with the grin, which remained some time after the rest of it had gone.

    "Well! I've often seen a cat without a grin," thought Alice; "but a grin without a cat! It's the most curious thing I ever saw in all my life!"

    --Lewis Carroll, *Alice's Adventures in Wonderland*, in M. Gardner, ed., *The Annotated Alice: Lewis Carroll, Alice's Adventures In Wonderland and Through the Looking Glass* (Middlesex, UK: Penguin, 1970), pp. 90-91.

    Gardner notes in the margin:

    The phrase "grin without a cat" is not a bad description of pure mathematics. Although mathematical theorems often can be usefully applied to the structure of the external world, the theorems themselves are abstractions that belong to another realm...

12. Serres, "Origin," p. 70. Serres' footnote 12 is in parentheses.

Illustration 6. Illustration of draftsman drawing a lute, from Albrecht Dürer's *Unterweisung der Messungmit dem Zirckel und Richt Scheyt, in Linien ebnen und gantzen Corporen* (Nuremberg: 1525). (Reproduced courtesy of the Rare Books Room, Pennsylvania State University Libraries.)

PART II

THE SEPARATE SCREEN

Illustration 7. Illustration from *Perspectiva pictorum et architectorum* by Andreae Putei (Andrea Pozzo), e societate jesu; from the original printed at Rome 1673 in Latin and Italian, translated into English by John James as *Rules and Examples of Perspective Proper for Painters and Architects* (London: Typis Benj. Motte, 1707). "On the Table A exactly squared, as you see here, fix the Paper, and the Rule B, having a Cross-stock serve as a Square, by the application of which STock to the Sides of the Table, you draw by the Rule what Parallel or Perpendicular Lines you have a mind to." (Reproduced courtesy of the Rare Books Room, Pennsylvania State University Libraries.)

Chapter 4

EDMUND HUSSERL'S ABSTRACTION

RATIONAL SPACE

More familiar than Proclus' idea of downward-projected images inhabiting the geometric imaginary, is that of geometric images as abstractions from objects in the sensible world. Edmund Husserl, more than Serres, gives us a compelling theorization of this process. His basic claim is that geometry (ideal praxis or formal abstracting) is initially a product of real praxis (natural abstracting).[1] To lay the ground for this claim, after indicating what he sees as the decisive character of modern science—the idea that "the infinite totality of what is in general is intrinsically a rational all-encompassing unity that can be mastered, without anything left over, by a corresponding universal science,"[2] that is, rigorous knowledge of the world without any residue—Husserl states:

> For Platonism, the real had a more or less perfect methexis in the ideal. This afforded ancient geometry possibilities of a primitive application to reality. [But] through Galileo's *mathematization of nature*, *nature itself* is idealized under the guidance of the new mathematics; nature itself becomes—to express it in a modern way—a mathematical manifold [*Mannigfaltigkeit*]. *What is the meaning of this mathematization of nature?* How do we reconstruct the train of thought which motivated it?[3]

Adopting for the moment David Carr's reading of the term "real" as referring "to the spatio-temporal world as conceived by physics,"[4] ancient geometry—fielded through the Platonic schema of a higher reality which gave voice to the idealization of such things as "numbers, units of measurement,...figures in space, points, lines, surfaces, bodies,"[5] coupled to the development of an axiomatic base or unquestioned first principles—was able to conduct its operations free from the vicissitudinous character of the perceptual world. Ancient geometry participated or shared (*methexis*) in the benefits derivable from the static world of unconditioned truths, the world of "pure rationality" that typified the Platonic higher reaches of intellection.[6] A total world open only upon itself, but capable of producing through apodictic arguments a consistent insight. It did this by forming a totally rational space separated from the empirical. However, this very enclosure, this wrapping itself in itself in a certain finitude in the form of Platonic degrees of adequacy (geometry as exemplar), seriously limited the "possibilities" of ancient geometry's "application to reality," for it set itself only "finite tasks...knows only...finitely closed a priori."[7]

In this model, geometry was a closed conceptual ladder from which one might rarely glimpse, almost "accidentally," the Platonic ideal world of unchanging idealities.[8] Thus, geometry's gaze was directed upward (or inward toward *nous*) and away (with the exception of Platonic cosmology) from the possibility of a systematic quantification of visible and mercurial nature. This limitation or self-enclosure could be expressed as geometry quâ geometry. More expansively, geometry unfolded its own finitude through the employment of itself in finite tasks designed to circle back through the portal of systematic deductions to the basic elements of its own system. Ancient geometry proved its own validity and certainty by constantly returning, in prescribed demonstrations, to the self-evidence of its own grounding axioms. In sum, geometry was a practice within prescribed boundaries expressed as propositions that demonstrated the all-pervasiveness and essentiality of a hand-full of self-evident axioms. This grounding of its practice in self-evident truths gave geometry the status of exemplar that approximately reflected, for each geometer who adhered to this program of rational precision, the unconditioned character and potentiality of *nous*. In simple terms, precision reflected purity-truth.

In contrast to this limitation of an inward gazing to a finite set of unquestioned truths, for modern geometers the task is quite different. However, modern geometry still owes a debt to the ancient space of a systematic coherent rationality:

> the possibility of the infinite task...[is] linked as a matter of course with the concept of geometrical space and with the concept of geometry as the science belonging to it. To ideal space belongs, for us, a universal, systematically coherent a priori, an infinite, and yet—in spite of its infinity—self-enclosed, coherent systematic theory which, proceeding from axiomatic concepts and propositions, permits the deductively univocal construction of any conceivable shape which can be drawn in space. What "exists" ideally in geometric space is univocally decided, in all its determinations, in advance. Our apodictic thinking, proceeding stepwise to infinity through concepts, propositions, inferences, proofs, only "discovers" what is already there, what in itself already exists in truth. What is new, unprecedented, is the conceiving of this idea of a rational infinite totality of being with a rational science systematically mastering it. An infinite world, here a world of idealities, is conceived, not as one whose objects become accessible to our knowledge singly, imperfectly, and as it were accidentally, but one that is attained by a rational systematically coherent method. In the infinite progression of this method, every object is ultimately attained according to its full being-in-itself [*nach seinem vollen An-sich-sein*].[9]

The ongoing productive character of this geometrical space, a production toward infinity, is a determined domain laid out in advance by the very nature of the logic mobilized. It is determined by the self-same logic constituting the prescribed method for all who wish to interrogate this space. The ideal forms comprising this space are in this sense an expression of this method. They, and those yet to be created, are situated firmly and always only within the bounds laid down by this self-same method. A method that is a coherent system that speaks of a coherent space. This is a self-enclosed, yet always expanding, world of idealities that agree with the method employed to produce these self-same idealities.

In a word, geometric idealities are an exact reflection of the method used to generate them. This is quite different from the Proclian model wherein geometricals merely reflected (imperfectly) the upper, unattainable idealities (partlessness) of *nous*. Conversely, nothing is left over in this modern geometrical space. No gap exists between method and idealities. Husserl will later contrast this with the free play characterizing his prescientific or natural imagination. The latter he will offer as the initial conditions (fundament) that produced geometrical (ideal) space.

For the moment however, he says that for the modern geometer, this notion of an infinite world being mastered by a coherent and systematic method was not restricted to the world of idealities. Features of the visible world were also brought in under the mathematical mantle by being transformed into the formal language of mathematics. Geometric figures were reduced to numbers and letters (algebra), as was movement, continua (calculus and the like). And, analytical or coordinate geometry made it possible to express mathematically specific points or objects in space. In this sense of an explicit expansion into the world of the empirical, the transformation of general features of the visible world into the formal language of mathematics, we may further understand what Husserl meant earlier when he spoke of a rational all-encompassing unity. That is, the promise of modern (Galilean) science is the construction (through mathematical reduction) of the world into a systematic, coherent formalization expressed as mathematical quantities. It is the construction of a new or separate (always complete yet always open) world of governing universal principles. The truth of the world at the level of its governing principles therefore emerges as calculable and lastingly universal.

Hence, this modern mathematical domain preserves in part its debt to the Platonic schema of an ideal world free from the vicissitudinous character of the empirical. Yet also, in a sense, it reaches out or back into this self-same mercurial world in the form of predictions generated from mathematical extrapolation. It generates a calculated opening-ever-outwardness from within its own rigorous coherent system of logic. However, there is need for caution in understanding the "obviousness" of the underlying universalities of the objects of the empirical world, in that, "prescientifically, in everyday sense-experience, the world is given in a subjectively relative way." From this experience we derive what for us individually counts as the world.[10] This world has for each and every one a specific ontic validity, or constitutes what counts as the ground from which, and in which, life has a specific meaning relative to our understanding of that world as our world. Although we are aware of each person's world as *different* in light of this individual counting or ontic validity, we do not think that there "are many worlds."[11] We understand precisely and "necessarily" that there is a common world (and the objects comprising it) out of which, and to which, it makes its appearance, and which count to each and every one in different or specific ways.

However, this belief in commonality raises the question of whether such a belief is of itself merely a necessary idea that does nothing more than ascribe an objective existence to things in themselves, or whether there is about all the objects we perceive something that constitutes in their individual appearances an actual, nonrelative truth. More precisely, whether there is, in every one of these appearances, "a content [which] we must ascribe to true nature?"[12] Husserl answers in the affirmative by saying that the legacy of "pure geometry" makes possible this belief precisely because of geometry's ability to "construct *idealiter*," "pure shapes" that give to these self-same shapes for us the "self-evidence of absolute, universal validity."[13] These pure shapes come to represent a set of forms that we can all agree are common (nonrelative) by virtue of them being able to be generated mathematically in a consistent and identical way. Shapes that also can be, in a manner of speaking, thought to be coextensive with the diverse appearances the objects of the world hold for each of us.

Hence, the legacy of geometry is that it allows us to *see* precisely these *forms*—forms that speak of a commonality in diversity but which leave in place this self-same diversity at the same time as they establish the existence of a common world. The promise of geometry in its common "pure" world and the multitude of our individual-diverse worlds are not necessarily contradictory precisely because the former is coextensive with all the latter. Thus the world of "pure idealities" retains its own specificity, yet at the same time is "constantly being practically applied to the world of sense-experience."[14]

However:

> so familiar to us is the shift between a priori theory and empirical inquiry in everyday life that we usually tend not to separate the space and the spatial shapes geometry talks about from the space and spatial shapes of experiential actuality, as if they were the same.[15]

To gain a sense of the significance that Husserl attaches to the latter point we need to follow his ongoing exposition of "experiential actuality."

LIMIT–SHAPES

In the "intuitively given surrounding world," we are able, apart from any mathematical schemata, to abstract from the objects open to our ordinary vision and derive "mere spatio-temporal shapes," which, as "not-geometrical-ideal bodies" conform approximately to the actual "bodies" we experience along with their actual content and so on.[16] However, regardless of how much we stretch or manipulate our imagination, the shapes we derive, although ideal in a sense, are not to be confused with the shapes of geometry. Because:

> No matter how arbitrarily we may transform these bodies in fantasy, the free and in a certain sense "ideal" possibilities we thus obtain are anything but geometrical-ideal possibilities: they are not the geometrically "pure" shapes which can be drawn in ideal space—"pure" bodies, "pure" straight lines, "pure" planes, "pure" figures, and the movements and deformations which occur in the "pure" figures.[17]

Hence:

> geometrical space does not mean anything like imaginable space or, generally speaking, the space of any arbitrarily imaginable (conceivable) world. Fantasy can transform sensible shapes only into other sensible shapes. Such shapes, in actuality or fantasy, are thinkable only in gradations: the more or less straight, flat, circular, etc.[18]

Moreover, "the things of the intuitively given surrounding world," are not static: they are always changing or fluctuating.[19] Therefore, their "identity with themselves" is only always approximate.[20] Because they are engulfed in the process of fluctuation, there is, as it were, no fixity in their apparent imperfection, and as always typical they constitute or construct their own gradient of themselves. Since they are always typical, it follows that they are always only approximations of themselves. It is in this sense of being always typical that, for Husserl, a gradient is formed whereby the abstracting from "things of the intuitively given world" and "the abstractly conceived shapes of empirically intuited bodies" always strive for a certain perfection. Although, he adds, these shapes do of course enjoy, for all practical purposes, a certain (historically contextual) acceptable "simple perfection" at any given moment. However, it is always "conceivable" that improvements can be made, thereby constructing the possibility of an always open horizon.[21]

What this amounts to saying is that these ordinary mental images, which are the product of an abstracting from the "intuitively given world" or the "empirically intuited bodies," are involved in a constant reformulation toward perfection: a "praxis of perfecting."[22] Furthermore:

> out of the praxis of perfecting, of freely pressing toward the horizons of conceivable perfecting "again and again," limit-shapes emerge toward which the particular series of perfectings tend, as toward invariant and never attainable poles...if we are interested in these ideal shapes and are consistently engaged in determining them and in constructing new ones out of those already determined, we are "geometers."[23]

So this process of freely perfecting is pushed-up, so to speak, toward the conception of limit-shapes. The latter investigation of these forms in their subsequent determined production constitutes the work of geometers. It is in this sense of a freely employed praxis, embedded in the everyday world, pushing-up the conception and possibility of limit-shapes and of a science that pertains to it, that Husserl wants us to understand his natural or real praxis and the world of its embedment as the fundament, originary condition, or initial force behind modern geometry (ideal praxis).

Husserl is describing something different from what we had earlier come to think of as the origin of the geometrical imagination. Recall Proclus' mathematical imagination with its less than pure imagery reflecting imperfectly the idealities of *nous*. Although it could be triggered by features of the empirical world, Proclus' imaginary comes into its full geometrical being only through a projection down from *nous*. Hence, geometricals reflect, but only imperfectly, the pure Being or partlessness of *nous*. They have little or nothing to do with the world of the sensible. Indeed, in this ancient schema, perceptual forms are argued to be twice removed from *true being* (that is, two layers down from the ideal forms of *nous*).

In contrast, for Husserl, it is his ordinary praxis of freely perfecting carried out in his nongeometric intuitively given world or from empirically intuited bodies (hereafter, the nongeometric imagination) which issues up, by its own processes of natural abstracting, the conception of limit or ideal shapes. This process replaces any projection down from a higher world of idealities. It is precisely because of this initial natural abstracting activity of the nongeometric imagination that there issued forth the possibility of the geometric space of ideal forms, leading eventually to the possibility of modern geometry.

The key point of difference is that Husserl has preserved the Proclian imagination which writes, but has altered the complexion of the written-upon aspect. He has abandoned the notion of *nous* writing upon the imagination to reveal itself, in favor of posing a symbiotic and necessary relation with empirically intuited bodies. The latter is one of the formative ingredients for the initial writing, or emergence, of the geometrical imagination and its products. The other formative ingredient is the intuitively given world. The Husserlian nongeometric imagination therefore freely writes toward the invariant, writes toward limit-shapes. It comes into being, or is initially written, from within (abstracting from the intuitively given world) and from without (abstracting from empirically intuited bodies), as much as it writes itself (the formation of "'pure' shapes" and the ongoing production of new ideal forms).

PRAXIS AND HORIZONS

However, Husserl sounds another caution. Apart from, or in place of, a "real praxis" freely considering the empirical possibilities of actual possible physical objects, what "we now have [is] an *ideal* praxis of 'pure thinking,'" one that "that remains exclusively within the realm of pure limit-shapes."[24] This ideal praxis of pure thinking therefore ultimately becomes its own fundament. It becomes its own source of generation, now entirely separated from its initial ground of the nongeometric imagination. Ideal praxis is thus a type of mathematical radical idealism or geometrical separatism, to use a contemporary phrasing.

Hence, there is an actual and distinct separation between these two worlds: a freely employed real praxis is separated from a formalized ideal praxis. The non-geometric imagination linked inescapably to the world of ordinary experiences is separated from a modern geometric imagination wrapped-up in its own world. To put this in a colloquial form: the ideal child has abandoned its parent and become a self-generative force within the enclosure of its own identity or praxis: *narcissistic measure*. Ideal praxis has disowned or disavowed its generative lineage. It has broken with its formative past. It has forgotten its own origins and any original meaning (in all its ontic fullness) pertaining thereto. Hence, this modern process of the production of closed-off idealities may be characterized as a process that hides, or screens from itself, its initial source or origin. It is a hiding, a screening or separation, that is enacted by the self-contained and self-generative character of ideal praxis (modern geometry).

However, if we take up this idea of a split imagination, the dual spaces of real and ideal praxis, what is not clear is precisely how we are to think this separation in its actuality. It is not clear how we are to think the gap between real and ideal praxis, because saying something is actually self-enclosed in relation to something else already presupposes what it is attempting to explain: separation. In order, then, to shed some light on this problem, let us again backtrack for a moment.

Husserl invoked the idea of the gradient to suggest a way to differentiate and separate both praxes. However, we cannot think the separation of these praxes as actual by any appeal to this feature unless we invoke the condition of a systematic scale of purity in relation to the imaging of shapes of each respective praxis. Such a solution would be a relative weak form of thinking actual separation, since it would constantly rely on a common scale upon which degrees of purity could be marked. With such a common scale, it is hard to imagine how these praxes could ever become actually separated in any distinct way. Indeed, although Husserl does say that these praxes can be actually separated from each other by a gradient marked, at one end, by naturally abstracted images and, at the other, by pure shapes, it needs to be asked: Is not the very idea of a scale or gradient of degrees of perfecting always already a geometric concept? Therefore, is not any possible differentiation of a respective praxes already contaminated by the very thing he wishes to differentiate, namely, systematized geometry? What sense of the non-geometric imagination may we imagine is possible from a *geometric* program of differentiation other than a more or less geometric one?

Of course, in Husser's lights, real praxis issued up initially in its processes of naturally abstracting the very possibility for the conception of a gradient, which was later to become marked or adopted by ideal praxis and transformed into a formalized scale (putting aside that the idea of a geometric gradient obviously existed in prescientific geometry). But, does this defense not confront us with a further dilemma by simply returning us, by way of a different route, to the problem of defining actual separation in terms of rigor? In this context, rigor is a term with decidedly geometric overtones. Indeed, in remaining faithful to Husserl, what definition of rigor in his gradient theory of actual separation is possible other than a geometric one? Concomitantly, his other idea of an open horizon or the possibility of an infinite forward projection also emerges as a feature of both praxes. First, in ideal praxis, this open horizon, which takes its lead from a determined ground, is understood as a desire to master a rational infinite totality of Being by the systematic employment of a rational science. As he says more clearly in a later passage: "the possibility emerges of producing constructively and univocally, through an a priori, all-encompassing systematic method, *all* possibly *conceivable* ideal shapes."[25] Certainly, his use of the terms "possibility" and "possibly" suggests that we are obliged to think of the ideal praxis as a praxis that can ideally or "*conceivably*" open-up to infinity from within the closure of its idealities or invariant limit-shapes.

This, of course, comes as no surprise, as we would expect the open horizon of an ideal praxis to be significantly ideal. Just as surely, in relation to his world of real praxis (nongeometrical imagination), there would seem to be little doubt that it also has a decidedly ideal character as well, notwithstanding its stated empirical associations. Husserl implies an ideal aspect of this real praxis as something that is *always* enfolded in its activity of approximation (that is, the constancy of an open horizon of improvements). The following passage supports this reading:

> Indeed, the things of the intuitively given surrounding world fluctuate, in general and in all their properties, in the sphere of the merely typical: their identity with themselves, their self-sameness and their temporally enduring sameness, is merely approximate, as is their likeness with other things....Something like this is true also of the abstractly conceived shapes of empirically intuited bodies and their relations. This gradualness can be characterized as that of greater or less perfection....[O]f course there is a limit to what can be done by means of the normal technical capacity of perfecting, e.g., the capacity to make the straight straighter and the flat flatter. But technology progresses along with mankind, and so does the interest in what is technically more refined: and the ideal of perfection is pushed further and further. Hence we always have an open horizon of *conceivable* improvements to be further pursued.[26]

In sum, it is not clear how we are to think, adequately, these praxes as uniquely and actually separate, given that they seem to share so many features. These features, however, are constantly being defined by processes we are led to understand are solely those of ideal praxis. In a word, the non-self-enclosed nongeometric imagination is defined by scaling mental depictions in terms of a gradient. To look at this from the other perspective; what we again see is that the imagination in toto has a *range* of cacographic utterances, some of which, due to their complex appearance, seem constantly to fluctuate freely to and from an unattainable norm or originary form of themselves (the natural abstracting processes of real praxis)—with the balance being largely repeatable due to their less complex or simple appearance (the formalized abstracting processes of ideal praxis).

Hence, geometry in its modern cloak emerges as a system that can endlessly repeat simple cacographic forms, yet still awaits (at the time of Husserl's writing) a system that can endlessly repeat complex ones. Moreover, because Husserl is driven to mobilize geometric concepts to substantiate and explicate the difference between his praxes, we might be led to conclude he breaks the borders of his own separation. He reaches out through the walls of his self-enclosed ideal praxis of *precise* open-endedness and brings his real praxis of imprecise open-endedness into the fold of modern geometry by defining difference geometrically. Due to this enfolding, ideal praxis therefore becomes the originary moment of real praxis. Indeed, only through the employment of specific geometrical concepts with which he characterizes his rigorous ideal praxis, may we now glimpse the imprecision or natural character of his originary real praxis. Thus, real praxis, in Husserl's model, is *geometrically* imprecise in its (nongeometrical) naturalness.

SEPARATION AND UNITY

However, it is also important to note that by mobilizing geometric concepts such as grades and rigor, Husserl allows us to think his praxes *as if* they were separate, albeit a separation that is geometric. Indeed, in this model, geometry emerges as a discourse *of* separation as much as it is a discourse seeking unity, seeking a rational totality. Mindful, it is something else again to imagine that either of these praxes can ever be thought *in* their separation, can ever be thought entirely separately in the finitude of some self-enclosed selfhood (anymore than we can ever think absolute unity or undifferentiated infinitude). Noting this difficulty, however, does not mean that we have lost the spirit of Husserl's earlier comment concerning the danger of thinking "the space and the spatial shapes geometry talks about" and "the space and spatial shapes of experiential actuality, as if they were one and the same." Quite the contrary. To note the contaminative effects of one in the other in any thinking of them as different is not then to think of them as the same or identical. It is simply to come to an understanding that one is constantly present as an absence in the other; that one is the condition of possibility for the other to be thought of as "other." More exactly, they construct from within themselves the possibility of the other as other to themselves by always absenting this other from themselves. This self is also constructed from within the other as "other" to itself.

This is precisely the work of geometry's binary created structure throughout its long tradition. This is its contaminative character. What is displaced in each pole of geometry's binary in order for each pole to be the other to each, is precisely this otherness of each that each in some manner or another declares is not itself. It is in this sense of a necessary interdependence and displacement that I mean to infer Husserl's real praxis is geometric in its nongeometricalness as much as his ideal praxis is nongeometrical in its geometricalness. One writes the other as much as it is written by the other. Geometry is, therefore, always already both ideal-empirical, regular-irregular, and at the self-same moment it offers itself to us as solely one: an ancient and modern geometry of regularity, or a contemporary geometry of irregularity. Indeed, geometry cannot hold the terrain free from the contaminative presence of its displacements, because we are constantly obliged to call back these unwanted others to think the privileged pole geometry wants to offer as its current terrain separated from these unwanted others. It is in this fashion that geometry constructs a horizon of immeasurability at the same moment as it constructs a horizon of measurability. Therefore, to think of geometry as a discourse of separation as much as it is a discourse of unity is to emphasize rather than deny the spirit of Husserl's insight: modern geometry seeks to separate itself from the abundant world of nature by the very act of defining or reducing features of this mercurial world into determined geometric terms.

Furthermore, at the very moment that Husserl gives us a way to think geometry as a rigorous determined system seeking separation through reduction, he also sets the conditions for us to speculate that modern geometry, under the Galilean umbrella, has not been able to explicate and take into custody all the abstracting ability of the imagination. Modern geometry has been unable to reduce all the imagery inhabiting the imagination solely to a rigorous formulation or a calculated repeatability. It has been unable fully to systematize its irregular companion, to geometricize *explicitly* the irregular in all its irregularity. It is precisely this limitation of modern geometry, the limits of reductionism, as we will see later, that constitutes the departure point from which contemporary geometry will take its leave.

Let us also not forget that Husserl's insights make it possible to understand very well the necessity of giving something other than a negative account of the empirical world or experience in relation to the activities of the imagination. Recall his idea of abstracting from the intuitively given world and from empirically intuited bodies. This is a natural process that freely strives repeatedly for perfection. What he does by noting this process of natural abstracting is to allow the empirical to seep up from below. Natural abstracting from the world of the empirical issues up the conception of limit-shapes and, in turn, acts as the ground from which the activities of ideal praxis initially take their leave. In so defining the features of his real praxis as a natural abstracting from the intuitively given and empirically intuited worlds, Husserl blurs, so to say, the traditional edges separating the images to be had in the imagination from those open to perception. In shorthand fashion, then, Husserl writes the empirical back into the equation. He writes it in as a formative presence beyond a mere trigger, in contrast, say, to the twice-removed empirical world of Proclus. Husserl also differs from Serres in this only insofar as he avoids the pitfall of ascribing non-meaning to the empirical.

Finally, for Husserl, the imagination is not an intermediate screen initially hovering between a higher world of intellection and the empirical world open to perception. Rather, it is from the free employment of the imagination (natural abstracting) that comes the possibility of conceiving the existence of a common world (limit-shapes) and the possibility of a rigorous space of endless productions (idealities). It is, therefore, only after the formation of this rigorous, reductive space that, for Husserl, the geometric imaginary functions as a screen, a hiding. This is a screen that hides as much as it reveals. It reveals a common world (limit-shapes) but does so at the price of forgetting its original ontic validity. Indeed, the explicit reductionist bent and self-generative processes of modern geometry must by definition lead inexorably to an effort to screen, hide, or cloak reality in its (geometry's) own methods. As Husserl himself says in one of his most celebrated insights (to paraphrase): geometry takes for true being what is just a method...it cloaks reality in a garb of ideas.[27]

But contrary to what he says about this observation (geometry's separation from the life-world), plus the forgetting of its heritage, and its screening or cloaking reality in a garb of ideas, we need to avoid jumping to his conclusion that the formalized practice of the geometric imaginary is free from being contaminated by what it has forgotten, screened, or cloaked. This is not so any more than it means that the free employment of the imagination, the nongeometric imaginary, can be theorized without showing the contaminative traces or stains of the geometric imaginary.

MEASURE AND METHOD

Before leaving Husserl, it will be useful as a preamble for the rest of this study to look more closely at this idea of cloaking reality in measure and method. This amounts to saying that we dress the world from the cupboard of the imaginary, itself always imagined in the style of the world we inhabit. To lay the ground for this idea of the cloak or covering screen of measure and method, Husserl turns his attention to "the concreteness of sensibly intuitable bodies." He tells us that the world open to ordinary everyday experiences is a world that exhibits "real-causal interdependencies." Moreover, changes to these "sensibly intuited bodies" (such as any change in their "spatio-temporal position" or in their form or in their general properties), are not arbitrary or accidental but "depend on one another in sensibly *typical* ways."[28] They may be observed to have their "habits," to "behave similarly under typically similar circumstances." Furthermore, the world as a whole also exhibits its "'habit,' that is, that of continuing habitually as it has up to now;" hence, "our empirically intuited surrounding world has an *empirical-overall style*."[29]

What Husserl is driving at here, is that in an everyday sense, the things of the world (and the world as a whole as it presents itself to us) have a certain discernible order or regulative pattern to their inter-relationships and activities. They sensibly hang together: "through a *universal causal regulation, all that is together in the world* has a universal immediate or mediate way of *belonging together*." Even when we seek to use our imaginations to conjure up a world other than that ordinarily open to us, or seek to "represent to ourselves the future course of the world," we still adhere in this imaginative construction to the "style" of the world as it has always appeared to us.[30] Moreover, by deliberately reflecting upon the world or by altering its possibilities by the free employment of our imaginations, we bring to consciousness, or make a subject of investigation the "invariant" character of the world. We bring to our awareness the "universal causal style of the intuitively given surrounding world."[31] However, all of this is very different from a scientific knowledge of the world. Out of the above we tend only to arrive at the general observation that events surrounding us at any given time or place appear to be causally determined. What eludes us in this general insight are the transtemporal, the "specifically determined world-causality, the specifically determined network of causal interdependencies that makes concrete all real events at all times."[32] Hence, what is necessary to construct scientific knowledge of the world, knowledge transcending the relative, the immediately experiential, is the devising of a method—one that affords us a way to construct in advance "the infinitude of causalities," and "compellingly verifying this construction in spite of the infinitude (of experience)."[33]

It is precisely here that mathematics comes to our service. For, as Husserl has told us earlier, mathematics is a space of idealities univocally determined, universally for all and capable of generating new idealities indefinitely. These features, coupled to "the art of measuring," bring mathematics down to earth, so to speak, by allowing it to *enter* the "empirically intuited world" of bodies and relate this world in an approximating fashion back to mathematics' own idealities:[34]

> All the things of the empirically intuitable world have, in accord with the world-style, a bodily character, are *res extensae*, are experienced in changeable collocations which, taken as a whole, have their total collocation; in these, particular bodies have their relative positions, etc. By means of pure mathematics and the practical art of measuring, one can produce, for everything in the world of bodies which is extended in this way, a completely new kind of inductive prediction; namely, one can "calculate" with compelling necessity, on the basis of given and measured events involving shapes, events which are unknown and were never accessible to direct measurement. Thus ideal geometry, estranged from the world, becomes "applied" geometry and thus becomes in a certain respect a general method of knowing the real.[35]

From this summation, Husserl asks:

> But does not this manner of objectifying, to be practiced on one abstract aspect of the world, give rise to the following thought and the conjectural question: *Must not something similar be possible for the concrete world as such?*...But is this not possible only if the method of measuring through approximations and constructive determinations extends to *all* real properties and real-causal relations of the intuitable world, to everything which is experienceable in particular experiences? But how can we do justice to this general anticipation...how can it become a practicable method for all concrete knowledge of nature?[36]

Can the world(s) open to the Subject (the world all of us experience in a relative fashion, experience in our own particular way with a corresponding ontic validity), can these worlds in all their relative manifestations be expressed in determined constructions similar to those of the idealities forming the structure of geometry? Can the experiential world be brought in under the manifold of the mathematical? Can it be brought in the same way that geometry, by applying its method of determined constructions grounded in idealities, has brought in the shape aspect of sensibly intuited bodies?

The difficulty that Husserl identifies as pertaining to this question is a formal one in the most limited sense of this term as relating to shape. The material plena cannot, "in their gradations, be directly treated as are the shapes themselves."[37] However, this is not to say that they must, therefore, be merely dropped-out of the equation. Rather, to sustain the sense of a science that can speak of the totality of the world, speak in its rational voice with nothing left over, as well as maintaining the sense and certainty of the world that it is a common world for all, these sensible qualities must in some manner gain an expression within the auspices of the mathematical mantle. Their certainty to us, running in an uninterrupted fashion throughout all Subjective interpretation:

> becomes attainable for our objective knowledge when those aspects which, like sensible qualities, are abstracted away in the pure mathematics of spatio-temporal form and its possible particular shapes, and are not themselves directly mathematizable, nevertheless become mathematizable *indirectly*.[38]

What, then, would a mathematization of the latter kind look like? What problems would it meet? One profound problem relates to exact and inexact magnitudes. For instance, the temperature of any object can be notated on some rigorous scale. However, its coldness or warmth is quite another matter. What is cold for one person is not necessarily cold for all. In other words, an object can be positioned on a scale of degrees, permitting us to say its temperature is exactly this or that. Nevertheless, this self-same scale does not issue-up for us a way to say with equivalent exactness, or in some universally acceptable sense, what the coldness of a specific object is. Such a scale does not allow us to abstract some feature of the object in question that will constitute the coldness for all. It does not allow us to abstract some ideality or universal structure of this or that quality. The way that a geometric shape, for instance, is thought to be an abstraction in relation to an object belonging to the empirical world. This is a shape that can be endlessly repeated by anyone in receipt of the formal procedures.

Thus:

> What constitutes "exactness"? Obviously, nothing other than what we have exposed above: empirical measuring with increasing precision, but under the guidance of a world of idealities or, rather, a world of certain particular ideal structures that can be correlated with given scales of measurement—such a world having been objectified in advance through idealization and construction. And now we can make the contrast clear in a word. We have not two but only *one* universal form of the world: not two but only *one geometry*, i.e., one of shapes, without having a second for plena.[39]

There simply is not a "mathematizable world-form" of plena in the way there is regarding shapes.[40] Qualities are not, in any analogous sense, idealizable. Consequently:

> the measurement ("assessing") of them cannot be related to corresponding idealities in a constructible world already objectivized into idealities. Accordingly, the concept of "approximation" has no meaning here analogous to that of the mathematizable sphere of shapes—the meaning of an objectifying achievement.[41]

It would be easy to overlook the significance of what is being implied here, due to the apparent obviousness—indeed, the almost trivial observation Husserl appears to make regarding the vicissitudinous and relative character of the qualities of things.

TWO GEOMETRIES

Husserl informs us that there is a geometry of shapes but not of qualities. However, without our needing to reactivate everything previously discussed, it can be seen that this notion of one geometry, a geometry only of shapes, is dependent upon adopting the traditional voice of geometry as the discourse that demonstrates the universality of the world. This is a discourse that captures, more or less, a common (and privileged) world in the form of austere figures, idealized principles, and so on. Of course, as far as it goes, there would be no point in denying geometry is, at least, certainly all of this. Yet, have we not also seen that geometry achieves this privileging process through a whole series of displacements? Have we not seen that it consistently marks out, or measures, the mercurial world by *taking this world's measure* (in the other sense of the term measure, as relating to a specific character thereof)? Have we not also seen traditional geometry always being very precise as to what will count as the mercurial world: its consistent interest in the character of the immeasurable, which it announces as radically irregular defined against its own regularity?

This being the case since the time of the Greek miracle of principled geometry, there has always been one measure enfolded in another. First, there is geometry constructing a common world in mathematical terms. Second, arising from within this measure, there is a *sotto voce*, an undertone, that speaks or *takes the measure of* the mercuriality of the world by defining this self-same mercuriality against the exactness of its first measure. The first, numerical measure, speaks of invariant forms that reflect the invariance logically demanded by the very conception of a common world. The second measure, that of taking the measure of—defining a specific character, demanding something measure-up, and so forth—consistently speaks of a variant character to the concrete world in all its visible operations. It is, therefore, in this double sense of measure that we have always had two geometries: the measure of presence and the measure of absence. One is enfolded in the other. More precisely, we have a discourse that measures very carefully, precisely, and exactly everything it says about the world of idealities by a process that *unmeasures* the other world it designates as outside its current purview. In taking the mercurial world's measure, it takes away its measure by defining its character as always variant, as being *directly* immeasurable. In this double sense of measure, immeasurability at this point is therefore another form of measure.

APPROXIMATIONS

The second matter of importance relates to Husserl's comments about approximation. Recall how Husserl stated that geometry's determined constructions at the level of the empirical (figures-measurements) can be correlated with—or, on a scale of exactness, made to approximate—geometry's "ideal structures." This idea is similar in principle to Proclus' figures in his geometrical imagination reflecting imperfectly the idealities of *nous*—except that Husserl has reversed the order of reference by having geometry at the level of the empirical reflect, approximately, the ideal structures of the geometric imaginary. Hence, the approximation he speaks of here is to be understood in the form of degrees of precision that advance steadily forward toward more and more perfection in measuring. On the other hand, as the qualitative configurations based on the sense qualities-plena are not analogues of spatio-temporal shapes, they cannot be related to corresponding idealities. This is so precisely because there are no idealities that can be constructed in advance as the universal form of respective sense qualities. Without this universal form, they have nothing to aim for. They have no desired goal to perfect. In other words, the uses of the term "approximate" only make sense if one knows in advance—or, more precisely, has already constructed in advance—a world of purity that something else can be said to correspond to, or can be said to be always striving toward. Hence, qualitative configurations are not approximations in the sense of this term referring to a scale of perfection moving toward ideal structures.

Harvey succinctly summarizes this point:

> Husserl notes that "a direct mathematization...in respect to the specifically sensible qualities of bodies is impossible in principle"....While there is an experience of gradations of sensed qualities of which we might assess the magnitude (say between bitterness and sweetness, roughness and smoothness, etc.), there is not an approximation towards exactness, which for Husserl is the distinguishing characteristic of mathematical articulation....Whereas mathematical idealities provide what Husserl calls "ideal limiting poles or shapes," intuition of sensory qualities occurs in relation to no such ideal limits. One almost always operates in pure intuition with the normative, comparative, and superlative—all of which have indeterminate degrees of variation.[42]

The significance of this relates to what we will later see when we meet the contemporary geometric imaginary. In this, the geometer, Benoit Mandelbrot, will simply drop out altogether the idea that his fractals (quantitatively or qualitatively) approximate anything. Instead, his fractals are *mirrors*. His argument in a preliminary form is: if his geometry can reproduce the image of nature open to ordinary perception, then, by definition, his geometry must *mirror* nature. Hence, his fractals approximate nothing. More grammatically, his fractals are not approximations. He gets rid of the concept of approximations by refiguring reflection as a truth-in-itself: an unproblematic one-to-one equivalence with the objects in nature. In this way, he seeks to bypass the problems which confronted modern geometry in its attempt to mathematize the plena.

INDIRECT MATHEMATIZATION

How, then, is the mathematization of the plena achieved in modern geometry? Let us examine Husserl's insight about indirect mathematization:

> Such mathematization is thinkable only in the sense that the specifically sensible qualities ("plena") that can be experienced in the intuited bodies are closely related in a quite peculiar and *regulated* way with the shapes that belong essentially to them. If we ask what is predetermined a priori by the universal world-form with its universal causality—that is, if we consult the invariant, general style of being to which the intuited world, in its unending change, adheres: on the one hand the form of space-time is predetermined, and everything that belongs to it a priori (before idealization), as encompassing all bodies in respect to shape. It is further predetermined that, in each case of real bodies, factual shapes require factual plena and vice versa; that, accordingly, *this* sort of general causality obtains, binding together aspects of a *concretum* which are only abstractly, not really, separable.[43]

Plenum qualities and shapes are necessarily bound together in a causal network apprehended as an always open horizon. Since geometry has already idealized the shape aspect of the world:

> everything which manifests itself as real through the specific sense-qualities must have its *mathematical index* in events belonging to the sphere of shapes—which is of course, already thought of as idealized—and there must arise from this the possibility of an *indirect* mathematization,...i.e., it must be possible...to construct *ex datis*, and thus to determine objectively, *all* events in the sphere of the plena.[44]

This idea of a general causality is, as Husserl notes, an idea enjoying a long tradition in one form or another. He gives the example of the Pythagorean correlation of pitch with the length of a vibrating string. As the string could be exactly measured, it followed that a mathematical scale could be produced expressing a causal relation between length and sound. Thus:

> all concrete intuitively given events in the familiar surrounding world contain easily discernible dependencies of plenum-occurrences on those of the sphere of shapes.[45]

However:

> not everything which concomitantly and visibly changed on the side of shapes was measurable through the old, developed measuring methods. Also, it was a long way from such experiences to the universal idea and hypothesis that all specifically qualitative events function as indices for precisely corresponding constellations and occurrences of shapes. But it was not so far for the men of the Renaissance, who were inclined to bold generalizations everywhere and in whom such exuberant hypotheses immediately found a receptive audience.[46]

Thus, mathematics came to be seen as:

> a realm of genuine objective knowledge (and technology under its direction)....There must be measuring methods for everything encompassed by geometry, the mathematics of shapes with its a priori ideality. And the whole concrete world must turn out to be a mathematizable and objective world if we pursue those individual experiences and actually measure everything about them which, according to the presuppositions, comes under applied geometry—that is, if we work out the appropriate method of measuring. If we do that, the sphere of the specifically qualitative occurrences must *also* be mathematized *directly*.[47]

What this comes down to is the idea that everything is immersed in a universal sea of causality—coupled to the possibility of a correlation between the plena and shapes expressed as mathematical indices—culminating (from Galileo on) in the general hypothesis the world could be dealt with methodologically. The task was to develop increasingly perfect ways of measuring. It was solely a question of method. Moreover, although the activity of measuring the data of experience is largely expressed as "inexact magnitudes and quantities,"[48] one must remember that measuring is itself also grounded in the idea of achieving more and more exactness, or of perfecting its own proficiency. Measuring is, simultaneously, a method of reducing elements of the world to quantification and a method for improving its own method.

Couple all this to what Husserl spoke of earlier regarding modern geometry's desire for the world to be systematically interrogated and encircled by a rational science with nothing left over, and the general hypothesis alluded to above may be comprehended or expressed in the following way: Geometry is a method explaining the causal relations of elements of the world (shapes and plena) through the art of measuring (directly and indirectly). At the same moment, it constantly seeks new horizons for its application. It also seeks to perfect itself toward greater and greater exactness in the measurements forming the basis of its application to those things it is able to bring within its domain of logic. Moreover, each step forward from a new development in this method is vindicated by a process of constant verification in the same mathematical terms grounding this self-same method. This constant verification does not, however, alter the fact the method always retains the characteristic of being a hypothesis. For the very conceptual ground of this method, its desire to encircle everything within its own rational domain, results in verification being also open-ended. There is always an ongoing process (of verification) of an always expanding horizon of application:

> In the unending progression of correct theories, individual theories characterized as "the natural science of a particular time," we have a progression of hypotheses which are in every respect hypotheses *and* verifications. In the progression there is growing perfection, and for all of natural science taken as a totality this means that it comes more and more to itself, to its "ultimate" true being, that it gives a better and better "representation" ["*Vorstellung*"] of what "true nature" is. But true nature does not lie in the infinite in the same way that a straight line does; even as an infinitely distant "pole" it is an infinity of theories and is thinkable only as verification; thus it is related to an infinite historical process of approximation.[49]

FUNDAMENT AND METHOD

However, and here we find another caution, the method which characterizes geometry has become substituted "for the only real world, the one that is actually given through perception, that is ever experienced and experienceable—our everyday life-world."[50] This substitution is possible precisely because geometry is able to construct such a convincing self-enclosed, self-generative world of idealities where everything is worked out in advance. This wholly rational space of constructed idealities also endlessly throws itself forward in predictions and widening applications.

However, this:

> inherited geometry, the inherited manner of "intuitive" conceptualizing, proving, constructing, was no longer original geometry: in this sort of "intuitiveness" it was already empty of meaning. Even ancient geometry was, in its way, τέχνη [technique] removed from the sources of truly immediate intuition and originally intuitive thinking, sources from which the so-called geometrical intuition, i.e., that which operates with idealities, has at first derived its meaning. The geometry of idealities was preceded by the practical art of surveying, which knew nothing of the world of idealities. Yet such a pregeometrical achievement was a meaning-fundament for the great invention of idealization; the latter encompassed the invention of the ideal word of geometry, or rather the methodology of the objectifying determination of idealities through the constructions which create "mathematical existence."[51]

Of special interest here, and in contrast to Serres, is that Husserl's initial conditions (for the emergence of geometric idealities) are not without meaning. They are not a non-meaning or noisy stratum that needs to be expelled in order for idealities to make their appearance. Quite the contrary. This fundament is a formative force precisely because of its meaning, a meaning that could possibly be reactivated in the sense of its original ontic validity (this reactivation is precisely the Husserlian phenomenological program). This reactivation will realign modern geometry with the peculiar character of its formative past, that is, with geometry as *geo-metria*, as that which measures the earth (in both senses of the term "measure"). Hence, Husserl's clever ruse in referring to ancient geometry as surveying, which in this context is meant to imply more than a mere numerical act. For to survey something is not only to measure it. It is also to *take its measure*, to look-it-over, so to speak. It is to look at that world (beyond its mere materiality, but without displacing this materiality) which is the necessary ground and horizon *to* which all sense of Being is directed and which it grows out of. It is in this full sense of the term "world" and the double sense of the term "measure" that Husserl understands the original ontic validity of geo-metria as pertaining to the earth.

However, as we learned earlier, these two meanings of measure are for Husserl now separated. This is precisely because the original set of a priori conditions that informs his originary fundament is not in its original form and ontic validity now present in modern geometry. It did not get passed down in the whole series of sedimentations marking the long tradition of modern geometry. Hence, it is not there to explicate, precisely because it is not a meaning couched in the formal logic language of geometry. Therefore, to stand any chance of reactivating the originary meaning of geometry and thereby realigning it with its initial task, we must look beyond geometry's sedimentations as they have been handed down. We must look toward their originary fundament to return to science its full ontic validity. We must return to science its full sense as a science *of* the world, in contrast to a (modern) science that always constructs in its own image (or logic) a pared-off world that comes to stand as *the* world. Quite simply, for Husserl, science quâ geometry has forgotten its original purpose:

> In geometrical and natural-scientific mathematization, in the open infinity of possible experiences, we measure the life-world—the world constantly given to us as actual in our concrete world-life—for a well-fitting *garb of ideas*, that of the so-called objectively scientific truths....Mathematics and mathematical science, as a garb of ideas, or the garb of symbols of the symbolic mathematical theories, encompasses everything which, for scientists and the educated generally, *represents* the life-world, *dresses it up* as "objectively actual and true" nature. It is through the garb of ideas that we take for *true being* what is actually a *method*.[52]

CODA

> It is because of the disguise of ideas that the true meaning of the method, the formulae, the "theories," remained unintelligible and, in naïve formation of the method, was *never* understood. Thus no one was ever made conscious of the radical problem of *how* this sort of naïveté actually became possible and is still possible as a living historical fact; how a method which is actually directed towards a goal, the systematic solution of an endless scientific task, and which continually achieves undoubted results, could ever grow up and be able to function usefully through the centuries when no one possessed a real understanding of the actual meaning and the internal necessity of such accomplishments. What was lacking, and what still is lacking, is actual self-evidence through which he who knows and accomplishes [the geometer] can give himself an account, not only of what he does that is new and what he works with, but also of the implications of meaning which are closed off through sedimentation or traditionalization, i.e., of the constant presuppositions of his [own] constructions, concepts, propositions, theories.[53]

There would be no point in denying that Husserl marks out a profound critical horizon and lays a serious challenge at the door of science. This challenge, I would venture to say, cannot be ignored or conveniently historicized by assuming that phenomenology is in some sort of intellectual twilight (even if it were, it still remains the most insightful program with which to acquire a critical perspective on science that is not predicated on a denial of the value of science). The challenge is this: it brings to the fore science's myopia. In this, Husserl points out an extraordinary feature of science. Namely, enfolded in science is an intellectual monstrosity: science is grounded in radical forgetfulness. In disavowing its theories as it goes along (required by the very idea of hypotheses-verification), it has lost sight of the very world (Husserl's full sense again) that announced its possibility in the first place. Less alarmingly, it has formulated a program for displacing those features of the world that it is unable to bring directly under its desired mantle.

Geometry is certainly, as Husserl alerts us, a discourse demanding through the rigor of its logical sedimentations to be taken at its (logical) word. But it is also a discourse seeking in its logical sedimentations to close-off the character of its own origins. Yet, it is always already doing more than just this. While Husserl is correct, it seems to me, when he suggests that geometry's sedimentations attempt to shut out so much of the world of the experiential (by demanding all value utterances about the world be in logical syntax), it is also equally correct to say that geometry has had quite a lot to say (in a devaluing way) about the mercurial world lying at any one time beyond this self-same logic.

Thus, the intellectual tragedy, if I dare use such a word, is not just geometry's myopia or its disavowal procedures. To be sure, these two are certainly tragedies of Greek proportions. But, it has also been consistently ignored by those in and out of science that the *how* of the way we have come to think of many aspects of the life-world arises or makes its appearance within the very discourse of geometry itself. It arises in measure as the *other* to measure.

In so doing, geometry, until recently, has chosen to privilege the austerely quantitative. It privileged what could be endlessly repeated. The truth of the world, in all its complexity and variation, became mere repetitions constantly readjusted to generate a parade of seemingly countless verifications of a single theme. In many senses, this program took the mercurial out of the mercurial, took the music out of music. Hence, the crisis in European science of the not too distant past was not just the forgetting of itself as a science of the world, a fact well documented by scholars since Husserl's ground-breaking insight. It is the continuing fact that the *true nature* science now wishes to privilege in the form of fractal geometry, was already constructed in advance as irregular, and has been so constructed for two thousand years. This is a nature that geometry played more than a passing role in constructing, and which it then displaced in order to privilege the unchanging character of idealities. This wayward and mercurial nature, deemed incalculable, was constructed in advance to be forgotten. Yet, its continual persistence in the domain of geometry—indeed, its necessary presence in sanctioning the above privilege—suggests that it was also constructed to be redeemed once a geometry of abundance arose.

Before engaging this redemption, two other peculiar matters concerning the geometric imaginary are worth briefly noting. The first offers a geometric imaginary absent of contents. The second proffers a geometric imaginary which has a particular content that has consistently been overlooked in the histories of geometry and of philosophy. This avoidance is especially interesting in the latter case since the site in question is arguably the most famous philosophic metaphor in the whole of Western culture.

NOTES

1. Edmund Husserl, *The Crisis of European Sciences and Transcendental Phenomenology: An Introduction to Phenomenological Philosophy*, trans. David Carr (Evanston, IL: Northwestern University Press, 1979).
2. Husserl, *Crisis*, p. 22.
3. Husserl, *Crisis*, p. 23.
4. Husserl, *Crisis*, p. 23 (nt. 1).
5. Husserl, *Crisis*, p. 21.
6. Husserl, *Crisis*, p. 21.
7. Husserl, *Crisis*, p. 21.
8. Husserl, *Crisis*, p. 22.
9. Husserl, *Crisis*, p. 22.
10. Husserl, *Crisis*, pp. 23-24.
11. Husserl, *Crisis*, p. 23.
12. Husserl, *Crisis*, p. 24.
13. Husserl, *Crisis*, p. 24.
14. Husserl, *Crisis*, p. 24.
15. Husserl, *Crisis*, p. 24.
16. Husserl, *Crisis*, p. 25.
17. Husserl, *Crisis*, p. 25.
18. Husserl, *Crisis*, p. 25. Carr's insert is in parentheses.
19. Husserl, *Crisis*, p. 25. Husserl's phrase here is: "fluctuate...in the sphere of the merely typical."
20. Husserl, *Crisis*, p. 25.
21. Husserl, *Crisis*, p. 25.
22. Husserl, *Crisis*, p. 26.
23. Husserl, *Crisis*, p. 26.
24. Husserl, *Crisis*, p. 26.
25. Husserl, *Crisis*, p. 27.
26. Husserl, *Crisis*, p. 25.
27. Husserl, *Crisis*, p. 51.
28. Husserl, *Crisis*, p. 30.
29. Husserl, *Crisis*, p. 31.
30. Husserl, *Crisis*, p. 31.
31. Husserl, *Crisis*, p. 31.
32. Husserl, *Crisis*, p. 32.
33. Husserl, *Crisis*, p. 32.
34. Husserl, *Crisis*, p. 32.
35. Husserl, *Crisis*, pp. 32-33.
36. Husserl, *Crisis*, p. 33.
37. Husserl, *Crisis*, p. 33.
38. Husserl, *Crisis*, p. 34.

39. Husserl, *Crisis*, p. 34.
40. Husserl, *Crisis*, p. 35.
41. Husserl, *Crisis*, p. 35.
42. Charles Harvey, *Husserl's Phenomenology and the Foundations of Natural Science* (Athens, OH: Ohio University Press, 1989), p. 17.
43. Husserl, *Crisis*, p. 35.
44. Husserl, *Crisis*, p. 37.
45. Husserl, *Crisis*, p. 37.
46. Husserl, *Crisis*, p. 38.
47. Husserl, *Crisis*, p. 38.
48. Husserl, *Crisis*, p. 40.
49. Husserl, *Crisis*, p. 42.
50. Husserl, *Crisis*, pp. 48-49.
51. Husserl, *Crisis*, p. 49.
52. Husserl, *Crisis*, p. 51.
53. Husserl, *Crisis*, p. 52.

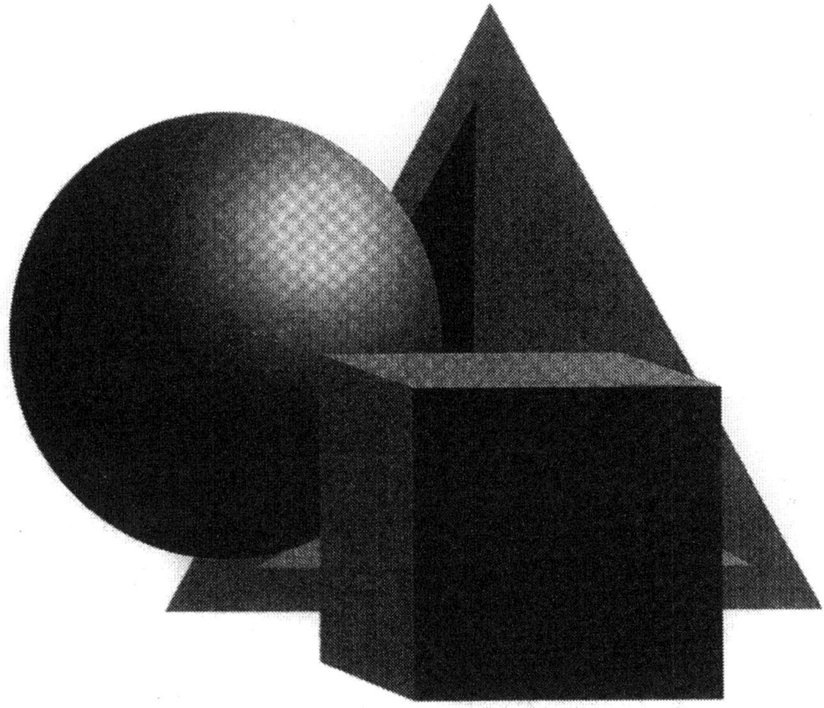

Illustration 8. Thought abstracting itself from its contents in computer-graphic space. Rendering by Ronald J. Avillion.

Chapter 5

CORNELIUS CASTORIADIS' SPACE

ABSTRACT SPACE

A recent theorization of the geometric imaginary, and one continuing the tradition of arguing for a separation between the lived-world and mathematical abstraction, is proposed by Cornelius Castoriadis. In a collection of essays on time, in which his contribution argues for the inseparability of time and Being, there is also an insightful discussion about abstract space:

> We can abstract, in thought, from that which is different, and think of pure difference as such. This is possible—and the result of this abstractive operation is pure, abstract space. In this space every point is different from any other point without any intrinsic characteristic—just by virtue of something external to it, that is its position "in" space. Two strictly identical cubes are different if and only if they are in different places in space. Abstract space is this miracle, this fantastic possibility of the difference of the identical. Points, equal segments, figures, or solids can be distinguished without any "proper" difference—because they differ as to location, as to their position in space.[1]

Castoriadis starts from the condition that abstract space is the product of thought abstracting itself from its contents. This means that abstracting can be either a process which constructs through reduction or a process which constructs through erasure, the latter being a radical reduction. The possibility of a pure difference thus emerges by expunging the contents of the geometrical imaginary. This abstract space gains its specific difference without recourse to any intrinsic characteristic. In so claiming, Castoriadis infers by default that ascribing an intrinsic characteristic is one of the processes by which we normally ascribe meaning to objects that do not necessarily fall under the mantle of the mathematical. Moreover, "extrinsic" and "intrinsic," used explicitly and implicity, respectively, become another way of saying that meaning functions in a relational field. It functions in this default instance by the play or relation of ascribed intrinsic characteristics—properties, qualities, and the like—of any ordinary object.

This implication of a relational field is carried over to abstract space. However, individual difference is here understood to be a product of an extrinsic ascription, that is, a specific position or location in this field. This postulation of difference in abstract space as thinkable because of spatial position, is an eloquent insight adding much to the time-honored idea that geometric shapes are formalized figures emptied of almost any content and hence are nonrelative. Castoriadis expands upon this old idea by introducing a relativity and tying it to position, that is, to where these figures are in space. This is a tie, however, which leaves in place, yet is also contrasted to, the apparent nonrelativeness or universality of these figures' shapes. Hence, geometric figures retain their nonrelativeness in terms of a formalized shape. But they also gain a relativeness in terms of their position in abstract space.

Still within the shadow of Husserl's schema, though of a deeper complexity and ultimately departing from it, Castoriadis explains the generative character of abstract space:

> Difference is infinitely productive: e.g., it underlies and makes possible the whole of mathematics. In mathematics we proceed by attaching characteristics to sets of "indifferent" elements, then making these same characteristics "indifferent" on another level, and so on. "Production" means building up from given elements and according to given laws. We can think of an infinite manifold of "identical" elements along one "dimension" or along any number of "dimensions"—and we have a "space-like" receptacle. We can fill this receptacle with objects produced as different—that is, reducible to each other and all to some elementary objects, according to determinate rules and laws: we then have a pseudo-physical, immobile universe. We can then add to it a supplementary dimension, call it time, and endow it with some peculiar properties, distinguishing it from the other dimensions of the pseudo-physical manifold.. . .We then have a four- (or n-) dimensional manifold, constructed from the identical and the different (that is, the identical repeated), and which we can think of and elaborate abstracting from any concrete content of it.[2]

Thus, in contrast to Husserl's idea, abstract space flows toward infinity because of a rationally coherent method that conceives a rational infinite totality (idealities). Castoriadis ties infinite production to the infinite number of positions, individuated levels, and dimensions possible in this abstract space. Specifically, starting from a set of nonrelative figures, themselves determinately generated from first principles or axioms, we then ascribe a location (an extrinsic characteristic) to some figure, thereby identifying any identical figures in an individuated way. A further differentiation is attainable by ascribing a level or dimension to whatever set of figures is under consideration. The figures in question can be considered "indifferent" to their counterparts on another level or dimension. Alternatively, we can think of them "along any number of dimensions." Infinite production is possible in this singular or multidimensional manifold, or abstract space, due to definite laws of procedure—laws that are given along with a specific set of first principles. Geometrical production is thus determined in advance, as indeed Husserl said before.

These procedures and possibilities of endless differentiations are carried out in a "space-like" receptacle. Being able to then fill "this receptacle with objects" suggests that Castoriadis is speaking here of the imagination constructing, or simulating in a fashion, the space open to general experience. In other words, the abstract space conceivable in the imagination is a receptacle-like container where (imaginary and identical) objects may be placed here or there, in much the same way as we can place objects inside a physical container. This idea is in accord with one of the most insightful comments about the imagination we have already met. Namely, no matter what world we may care to imagine, such imaginings are always in the style of the world we inhabit. However, we have to slow down our examination here, for it would be easy to overlook something new that Castoriadis is introducing in contrast to previous geometric imaginaries.

EMPTY SPACE

What Castoriadis puts into the equation is abstract space that can be thought devoid of content, thus implying the possibility of an actual separation between the two. He makes this point more explicit in a later passage:

> We...have a pure receptacle which can or cannot be filled with indifferent elements....Difference and abstract space are solidary; but they are extrinsic to that which is, in each case, different: e.g., two points. Thus, we can think of abstract space, precisely abstracting from any particular content: mathematics.[3]

The interdependence of difference and abstract space implied by the term "solidary" is ruled out of court by the reference to their extrinsicness. So, abstract space comes into being by abstracting (in thought) from differences, which results in pure difference. The latter is abstract space. In such a context, this is the same as saying that the imaginary can think itself without its contents; as Castoriadis says a bit later: "an empty space is both a legitimate mathematical concept and a possibility of our intuition."[4] Where it gets complicated is when we ask: In what manner can this pure abstract space be thought? Castoriadis has informed us previously it can be thought of as "a space-like receptacle." It is to be thought in the form of a receptacle that can be filled or left empty. Borrowing Serres' metaphor of the blackboard: all the geometric figures drawn on the blackboard can be erased leaving the blackboard still there. Hence, the conception or possibility of abstract space or pure difference is not dependent upon its contents. The blackboard is not dependent upon what can be drawn upon it.

However, the complication deepens when a further question is raised: What physical receptacle, upon which the imagination must draw to think itself devoid of content, is beyond the reductive purview of geometry? Posing this receptacle (at the level of the imaginary) as simply a convenient metaphor ("space-like") does not avoid this issue. Rather, it merely forces a coherency problem onto Castoriadis' schema.

To separate the imaginary from its contents, he must rely on giving some actual shape to the imaginary that can then be emptied. Without wishing to belabor the point, this shape-giving tendency cannot be avoided if we sanction Husserl's insight about the inevitability of all imaginings being *always* in the style of the world we inhabit.

The giving of some shape to the imaginary in the form of a "space-like receptacle" therefore calls up a certain finitude, which, of course, must then call up geometry. But geometry, in the form of the shapes that inhabit the imaginary, has already been expunged to make possible an imaginary space empty of shapes. The geometric imaginary is thus empty of geometry, yet as a space-like receptacle, it is still itself reducible to geometry. This is the same as saying that geometry is, first, the condition for thinking the geometric imaginary with contents and, second, is also the condition for thinking the imaginary as a space-like receptacle empty of contents, empty of geometry.

To ease this tension, the only conclusion possible is that geometric processes and their products (the imaginary and its contents) are inseparable in any absolute way since geometry is the condition for initially thinking their separation. Strictly speaking, abstract space is never pure (empty of its contents) for its "purity" can only be thought in the manner of its contents (thought geometrically as a space-like receptacle). These contents, by definition, are not of themselves pure, or their expulsion would not be logically necessary in order for the "purity" of abstract space to be thought in the first place. Since this purity of abstract space is tied to emptiness, we must assume that its initial contents are not of themselves partaking of this selfsame purity. To put this differently: were each content a small pure abstract space in its own right (a homeomorphism), then emptying abstract space to make it pure would be incoherent. Thus, Castoriadis' abstract space, his geometric screen that erases its contents, cannot be thought of as pure. Due to the inexorable link between the geometry of its contents and the geometry of the finite shape, he is forced to give his imaginary in the form of a receptacle.

Even if we were to grant the possibility of an expulsion of the imaginary's contents, we would simply be, in a round-about fashion, reinforcing what we have already noted when speaking about cacography. Namely, because the necessity for expulsion depends on these contents being to some degree not "pure spaces" themselves, the very idea of geometry as solely the science of purity is something of a chimera. In sum, Castoriadis' screen of the imaginary is initially stained by impure contents, and later, assuming the possibility of his emptied abstract space, also stained by the self-same imaginary being thought in the style of its previously expelled impure contents.

IDENTICAL AND DIFFERENT

Finally, Castoriadis' miracle that "pure, abstract space" is the "fantastic possibility of the difference of the identical" is a miracle that must also own to a certain impurity, because it is not only position that defines the difference of the identical in abstract space. Each and every geometric figure that is imagined as occupying the imaginary is itself different by virtue of being imagined in the first place. *To imagine is always to imagine differently.* Which is to say that the *identical* in all its instances is forever cacographic. It is forever open to the mental pathology of those who think or imagine. Hence, the miracle is not just that this *identical* may be thought as different (for it is an *identical* that is always different due to the ongoing difference of all subsequent imaginings). But the miracle is that differences can also be thought of as identical. Abstract or imagined space is therefore always the fantastic possibility of the identical that is never quite identical yet can be imagined *as* identical. It is always the fantastic possibility of imaginings as always just this or that way differing from themselves yet thought under the mantle of being *identical*—which then goes on to be thought (again) as different, due to position, and so forth. (Husserl's "common" and "relative" worlds gain another expression here.)

Moreover, within this schema countless differences are subsumed, first under the idea of the identical, and then under the rubric of *one* difference (that of position). Indeed, this is the legacy of geometry, of geometric idealities, of the geometric imaginary: the relative ease by which simple cacographic forms (traditional geometric figures) can be generated in the imaginary permits us to imagine the very idea of the identical. This is an idea that then loops back upon itself and seeks to displace the very difference that initiated its birth by subsuming all these differences under the rubric of one difference (position). Thus, difference is the condition of possibility for the identical to be thought as that which will unify difference. This is the whole legacy of geometry. It differentiates *through measure* what it then seeks to *unify* through the self-same measure—which it then seeks to reemploy as the difference that will count as *the* difference.

To stress this point: it is not that these austere figures can be repeatedly imagined as *exactly* the same. To adopt this view unproblematically would be to confuse repetition with exactitude. It would be to argue that the products of the imaginary are invariant (the death of creativity?), which in turn would be to argue for some actual separation between process and product. Nor is it a matter of perfecting toward some limit shape (if by this we mean a shape solely dependent upon simplicity). Rather, limit-shapes are a process relating to repetition. *Any* shape can be a limit shape if we can repeatedly bring it forth in the imaginary *in its own style.* Even complex shapes can be repeatedly imagined—it is just more difficult. This is a point that artists, musicians, and chess players (and geometers interested in plus-three-dimensional shapes) know all too well. Hence, we can repeatedly imagine simple figures, imagine them as identical, simply because we can easily imagine them in their own style. The fact that all these countless imaginings are never *exactly* identical with themselves has been something passed over by traditional and modern geometry in their haste to privilege the ease of repetition in contrast to determining the subtle differences evident in the products of this repetitive process. And speaking of overlooking, let us for a moment turn our attention back to an ancient gaze.

PLATO'S POOL

Plato's pool, or watery reflective screen, stands at the very dawn of Western geometry. This is a screen to which we have already given a cursory look, but one to which we need to return in more detail, not least because the long history of geometry has never been very far from the edge of this pool. Indeed, geometry's extraordinary story, in all its complex finitude, dances across the surface of this pool in the form of the images reflected therein. Freezing this pool's surface for the duration of this study so we might more easily see the traces of history is precisely why I have used the metaphor of the dancer leaving traces on the glazier. Freezing one metaphor to produce another is, of course, also a shorthand way to speak of the history of geometry and its privileging processes. So, in this cold light of the dawn of the Greek miracle, let us now take another look into this pool.

In Book VII of *The Republic*, Plato speaks of the prisoner's journey from the cave of shadows:

> And suppose once more, that he is reluctantly dragged up a steep and rugged ascent, and held fast until he is forced into the presence of the sun himself, do you not think that he will be pained and irritated, and when he approaches the light he will have his eyes dazzled, and will not be able to see any of the realities which are now affirmed to be true....He will require to grow accustomed to the sight of the upper world. And at first he will see the shadows best, *next the reflections of men and other objects in the water*, and then the objects themselves; then he will gaze upon the light of the moon and the stars; and he will see the sky and the stars by night better than the sun or the light of the sun by day?...And at last he will be able to see the sun, and not mere reflections of him in the water, but he will see him as he is in his own proper place, and not in another; and he will contemplate his nature.[5]

In a traditional reading, the reflections of heavenly bodies are seen as a set of second-order truths that act, in their watered-down or austere state, as a preparatory device. They reflect in a once-removed fashion the brilliance or power of a higher truth. Their austerity sheds (reflects) light on the realm of universals, on unchanging idealities. The lack of apparent three-dimensional materiality of the reflected figures further reenforces the idea that these figures are emptied of any content, that is, subjective bias. Moreover, the reflective surface of the pool masks the dimensionality of the pool itself. The pool is only its surface. The prisoner sees only the images on the surface. The pool therefore acts as a reflective screen interposed between the prisoner and a higher truth (portrayed in this story as the brilliance of the sun).

What we have here is the altogether quintessential image of traditional geometry's task. The attainment—or at the very least, the glimpsing—of lasting truths through the employment of austere figures that are thought to reflect, approximately, the purity of idealities. Metaphoric transference allows us to think of the reflections of planets as circles, and the reflections of distant stars as points. This ancient screen of science thus announces a discourse that mediates, filters, reflects, reveals, the truth of nature. It reflects on the screen of its practice that which is lasting and universal.

The other feature of this pool is the recursiveness of its images. Day after day, the images can be seen. True, depending on the time of day, they differ as to where they might appear on the surface (what better image to grasp Castoriadis' eloquent insight of the difference of the identical). Equally true, the prisoner would know that they are not really the same images as on the day before. For with every rising and setting of the sun, these images come and then completely go from the surface of the pool. Yet, every day they reappear, and because of their simple outlines (made simple by the great distance from the pool of the source object), they appear to be the same images as on the day before (hence, difference as identical). It is little wonder that this metaphor has exerted such an extraordinary effect on the history of science and philosophy. It is hard to imagine one of greater elegance.

In this graphic story, we also have an apt image with which to think Proclus' geometric imaginary. The always reflective surface of the pool, coupled to the persistence of the orbits of the heavenly bodies, may be seen as reflecting the idea of the self-generation of Proclus' geometric imaginary. The pool repeatedly reflects. It ceaselessly reveals a higher reality. Moreover—and this is the point I wish to stress—this continuity of repeated reflections, always in the style of their previous reflections, may be seen to illustrate the ease of thinking different instances of simple figures as *identical*. They can be thought of as identical even with the full knowledge that they are not quite the same images as on the day before.

Yet, austere figures are not all that is reflected in this pool. The prisoner also sees the "reflections of men." This particular image nevertheless seems to play no further overt part in the past discourse of geometry in its seemingly endless construction of the revelatory screen of the geometrical imaginary and its contents. It is, we could say, a forgotten image. It has fled. It is a lost reflection. In another manner, we could ask: Why are we not justified in thinking of the "reflections of men" as figures on a par with the planetary reflections, on a par with traditional geometric figures? Why do these "reflections of men" not stand in the gallery with the rest of traditional geometry's images, associated with the exemplary way?

Proclus, for one, would say such a reflection does not reflect any essentiality of *nous*. It therefore has no place in the geometric imaginary. For him, if this reflection was of any significance, it would be merely that of the seat, or home, so to speak, of the imaginary. A reflection of the material vehicle ("men") wherein the imaginary resides. However, to retain consistency in his schema, it would not be logically proper to reflect in his imaginary that in which the imaginary was itself contained. Best to just forget this image, which is precisely what Proclus did. For Plato, by contrast, such a reflection would be merely another version of the shadows projected on the wall of the cave which the prisoner had previously mistaken for reality. After all, it is the very character of that which transcends specificity, transcends this or that *man* at the same time as it resides in all *men*, that Plato's whole schema is intended to show. Therefore, the significance, if any, of these reflections being in this pool would be to remind the prisoner of the task ahead, namely, that of seeing, beyond the outer appearances of the self, into the very arcana of *men*.

Serres would almost certainly regard this reflection as visual noise—necessary, but in need of being overlooked in order to hear the sound of the spheres, needing to be overlooked in order for the heavenly reflections to communicate their message to the prisoner. Since I did not discuss Castoriadis' wonderful insights into the relation of space and Being, it therefore would be inappropriate to speculate on what he would make of these reflections. Suffice to say that I imagine he would agree in principle with Husserl. Speaking of whom, Husserl would almost certainly say that such an image could not be geometrically schematized in any way that would determine in advance its repeated appearance. Such an image is imbued with too much complexity, too much subjective bias, ever to fall under the mantle of the mathematical. This image is simply too full of content to be regarded as a limit shape. However, this notion of there being always already a fullness to this particular image gives us a glimpse as to the reason behind its erasure from the gallery of traditional geometric figures.

The promise of "seeing" the *full content of man* (the idealities that reside in upper intellection) is what the Platonic exemplary way is initially mobilized to fulfill. Hence, to pledge and fulfill this promise, we cannot at the same time see this image of the "reflections of men" as a traditional geometric figure (austere, empty of content) and yet also in some manner as pointing toward an entity (man) full of content. Which is to say that we should not lose sight of who is looking into the pool. For the entity who the image in the pool is a reflection of, is also the entity who causes the reflection in the first place, is also the entity who now sees itself in the reflection of itself. It is an image of a self seeing itself as the self it knows itself to be, which is a knowing of itself as a self full of content (regardless of whether it may know what this content is in all its fullness). Hence, in this ancient schema, what is transcendent in *man* cannot be an image, because images belong to the lower two realms of the sensible (the imaginary and the senses). Of course, in the imaginary, images can be employed to point toward this transcendent. But such images must first be of the type that appear empty of content and that themselves display little or no apparent change in their forms (austere figures quâ reflections of heavenly bodies). The "reflections of men" clearly do not fulfill this requirement.

Alternatively, with the source object of this "reflections of men" being so close to the pool (the prisoner and others looking into the pool), the differences that constantly plague their outer appearance would clearly be visible each day. Each reflection would always be noticeably different, yet still persistently full of content, whereas the reflections of planets and stars always look the same, because their differences are significantly reduced by the distance from the pool of the initial heavenly bodies that are their source. Conversely, the persistence of visible difference is the peculiar and unique character of the "reflections of men" and that of which they are reflections.

If such is the case, then in looking into the pool and repeatedly seeing his own reflection there, would the prisoner primarily note the differences or the samenesses of his own reflection? Is this what Narcissus, for one, is looking for? To see difference, to see change in its changing? Or to see its arrest? Of course, seeing difference arrested might be possible, but only if the pool gazer never took his eyes off his reflection. Only then would his reflection always appear the same to him (due to the fact that change to one's appearance is largely imperceptible on a minute-by-minute basis), even though he would also know he must be changing with the passage of time (the folly of the self-gaze announced in the facility of memory being marred by knowledge). More important for our concerns, would Narcissus ever be able to look at his own reflection and see it as empty of content, as he would with those of the heavenly reflections? Would this not be the death of Narcissus, death of the self gazing self? Would it not be the death of the whole Platonic project, as it relates to a "seeing" of the static *nous* that informs the upper intellection of each and everyone?

Given, then, the possibility of this reflection causing such a ripple or tear in the Platonic fabric, it is better to forget, to let this image flee from the discourse of traditional geometry. For even though such an image is indeed a limit-shape because it is recursive, it nevertheless reappears each day as an image that images its differences (as much as any samenesses) from those of the day before. This of course is also true of the heavenly shapes. But—and this is the key to the history of geometry and philosophy's forgetfulness—the differences in the heavenly reflections are less visible or obvious, unlike the "reflections of men" being obviously never quite the same as the previous day.

IDEALITY OF DIFFERENCE

So, in these "reflections of men," it is difference that is figured, not sameness. It is, therefore, an image that constantly *imperfects* itself toward a limit-shape of difference, rather than an image that constantly perfects itself toward a limit-shape of sameness. Which is to say, the preeminent style of itself is one of difference over sameness. Hence, being always reflected in its own style, it must always reflect difference in comparison to the seeming unchanging heavenly reflections. Moreover, these "reflections of men" are the complex *limit-shapes* of the entity-that-reflects, reflections of the living entity who is reflective. As such, and within the metaphoric sense of this pool of geometry, any inclusion of this image in geometry's gallery would have to be as an image of the *ideality of difference*. It is an ideality thought in the manner of a complexity, the always-changing character of change, the ever-differing of difference.

If this is so, then limit-shapes that are complex (constantly irregular) in appearance can be recursive in the geometric imaginary but are more difficult to formalize precisely because of their complexity, their irregularity. This difficulty is augmented by trying to think, repeatedly, the constant variance of the variant. To be sure, little effort is required to produce a reflection of this constant variance in Plato's pool. One merely looks into it. However, to then transpose this relative ease into a formulized feature of the geometric imaginary is quite another thing. Indeed, it is precisely the difficulty of this transposition that has made these "reflections of men" lost reflections in traditional geometry. From the point of view of this geometry, these "reflections of men" are properly the purview of the arts.

Indeed, this is precisely where Plato relocates them, in a fashion already alluded to. They are inside the cave in the form of mercurial shadows projected onto the back wall. It is these shadows (dark reflections requiring only a made representation of the sun: a fire) to which Plato draws the reader's attention in preference to those of a similar nature in the pool. The source of these internal shadows are figures who wander in the semi-light carrying art objects above their heads. It is therefore here, inside the cave and far from the pool of geometry, that variance is thus imaged, and always recursively. And these projected shadows in their *impenetrable* surfaces are more or less empty of content. Making them shadows, dark reflections, images of the overall deficient reality of living in the dark, is certainly a clever way to diminish or displace the obvious visible content of the similar figures which later appear in the pool. At the same time, this later appearance tells us that Plato thinks *man* is full of content and that the shadows in the cave do not reflect this fact.

This is the main point of the story. The prisoner takes for reality what are only empty (quâ impure-irregular) shadows. Hence, what is needed to raise the reality consciousness of the prisoner is another set of empty figures that are not merely shadows, not dependent upon the darker and half-flickering light of a fire, a false sun (representation as lie?). These new figures are the geometricals inhabiting the pool. These new figures are in the light. They bring the light. They bring into the light what was previously lost in shadows. That they are, in their turn, like the shadows, empty of content, is not of itself an incoherency. For these new figures are not reflections of some specific human being. Hence, in their nonhumanness, these geometricals will always be the same figures for everyone, as will the apparent continuity of their sameness (their simple outlines). Thus, in relating to no one in particular, they relate to everyone. They are the exemplary way made manifest, more or less, at the level of the visual.

Yet, because of what we have seen in the previous part of this study, and because we have now seen for ourselves these forgotten "reflections of men," these lost reflections continue to exert a curious pull even in the shadow of geometry's attempt at purity. This is not least because irregularity creeps back in the form of cacography, of mental pathologies. Indeed, irregularity (the "diseased" signature of the entity-who-reflects) is precisely the very thing that traditional geometry must employ repeatedly to privilege the exemplary character of the austere reflections in this pool of its practice. It is precisely because of the necessity of this employment in the service of privilege that geometry must always return (under a certain cloak of darkness) to these lost reflections.

For, as we have seen, these "reflections of men," reflections of difference, of irregularity, make their presence felt throughout the history of geometry in the form of the pathology of imaginings and in the form of the irregularity of graphics. They make their return as pathological stains that austerity can never quite wash off. They are stains or traces that mark the point from whence geometry wends it way toward a rational utopia. And they mark the point to where it must return repeatedly in order always to privilege the reflections of the heavens, to privilege geometry's horizon of measurability. These lost reflections thus return, repeatedly, to haunt any easy acceptance of geometry's past history as being the history of purity.

CODA

Returning to Castoriadis, at the same time as retaining a certain emphasis on Plato's pool as a contaminative screen that joins as much as it separates the regular from the irregular, the austere from the abundant, we may note one more important point as a propaedeutic to Part III. In Castoriadis' theorization of the imaginary as a pure abstract space, it emerged as a static site whereupon the production of the geometricals took place. However, although he argued that it could be thought devoid of its contents, there was no sense that it itself was generative, say, in the style of Proclus' imaginary. Indeed, it was the geometricals, the possibility of the difference of the identical, that captured his interest, over and above their initial generation as geometricals. What generative force was alluded to was mainly in the form of a vague power informing the geometricals' positions within this abstract space. Sometimes this power was suggested to be an abstracting process of thought, at other times, a set of determined rules of operation.

In contrast to this, what we will see immediately is geometry shifting its privilege from the staticness of idealities to the mercurialness of generation. This shift announces the moment when the geometric imaginary leaves its traditional home inside the mind of the geometer and ventures out into the world. It announces the moment when the geometric imaginary becomes, explicitly, a material screen upon which nature *itself* (in contrast to the inner-nature of *humanity*) takes over the dance. In the geometric imaginary openly courting impurity, the faint traces we have been attempting to see now leap into full view. Indeed, they gain a luminosity, brilliantly lit as they, in their turn, now seek to cast into shadows that which for two thousand years stood in the light: geometry's idealities. Plato's imaginary pool, which I emphasize, repeatedly, stands at the dawn of the Greek miracle of purity, is now finally made manifest at the level of the material. But, like all manifestations of ethereal dreams, its new material reality bears the stains of its irreal heritage. The traces may now be far from faint, the dancers may have changed places, but the dance remains the same.

NOTES

1. Cornelius Castoriadis, "Time and Creation," in *Chronotypes: The Construction of Time*, eds. John Bender and David Wellbery (Stanford, CA: Stanford University Press, 1991), pp. 56-57.
2. Castoriadis, "Time and Creation," pp. 56-57.
3. Castoriadis, "Time and Creation," pp. 56-57.
4. Castoriadis, "Time and Creation," p. 58.
5. *The Dialogues of Plato*, 2nd ed., trans. Benjamin Jowett (Oxford: Clarendon Press, 1875), pp. 401-402. My italics.

Illustration 9. Illustration from *L'architettura di M. Vitruvio Pollione*, Italian translation and commentary by Marchese Berardo Galiani (Naples: Stamperia Simoniana, 1758). The quotation from Vitruvius, *De architectura*, book I, chapter 1, [para. 11,] may be rendered as follows: "In order for so great a profession as this to be adorned by, and abound in, varied and numerous accomplishments, I think that only those persons may directly and justly claim to be architects who from childhood have climbed the steps of these disciplines and been nourished by most of the literary and artistic sciences have reached the highest temple of architecture." (Reproduced courtesy of the Rare Books Room, Pennsylvania State University Libraries.)

PART III

THE ABUNDANT SCREEN

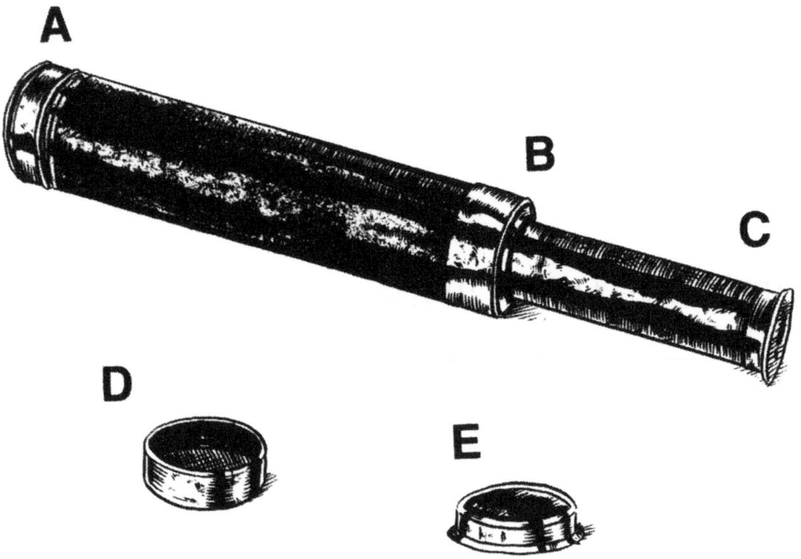

Illustration 10. Exploded view of David Brewster's two draw telescopic kaleidoscope, as manufactured by William Harris & Co. Opticians. According to the "Method for using the admired kaleidoscope, for creating an infinite variety of beautiful forms" instruction sheet that came with the instrument: "This Instrument generally consists of two tubes A B and B C one of which, B C is moveable within the other, C being end next the eye, and A the end next the object. The object cell D is made to screw in the collar A, as is also the lens E. In using the Instrument, place the object cell D in the collar A as far as it can go, push in the tube B C *completely,* holding the Instrument so that the sharp joint of the triangular aperture at C is downwards. The Instrument being now kept steady in the hand, turn round the collar A and by placing the eye at C will be seen a succession of the most beautiful forms and patterns." (Drawing by James Frazier.)

Chapter 6

DAVID BREWSTER AND THE KALEIDOSCOPE

SCIENCE OF THE BEAUTIFUL

The kaleidoscope is a primary metaphor for a particular conception of modernity. Its constellation of myriad images in jumbled disarray is a sign for the fragmentation of modern life and the distortion or loss of reality; the ever-changing images generated by this appealing apparatus serve to reflect the constant and rapid change of modernity. From Lewis Carroll's *Alice* stories to James Joyce's *Finnegan's Wake*, we readily witness the employment of the metaphoric energy of the kaleidoscope as a way to summon up a certain idea of change, as disruption and fracturing. In fact, the kaleidoscope—like the telescope, the microscope, the camera, and the computer—has demonstrated a capacity for metaphoric transportation across a wide range of disciplines.

Notwithstanding this metaphoric presence in almost every nook and cranny of our cultural practices and productions, there is a marked paucity of theoretical comment about the kaleidoscope. The only substantive and extensive discussion having a philosophical air, so to speak, remains that of its inventor, David Brewster. Against the backdrop of the current interest in the everyday, the theoretical poverty of something so familiar raises the question: How is the kaleidoscope to be thought, given its obvious theoretical transparency? How is this philosophical toy (a term favored by Brewster) to be thought beyond its historical displacement into the world of the nursery?

Brewster himself gives us a lead:

> The name Kaleidoscope, which I have given to a new Optical Instrument, for creating and exhibiting beautiful forms, is derived from the Greek words χαλος *beautiful*; ειδος *a form*; σχοπεω, *to see.*[1]

Enfolded in the unequivocality of this definition rises a further question: Can there be a science of the beautiful? More specifically—and the relevance of this will become apparent later—how are we to think a geometry of experience?

In 1814 (in an experiment on the polarization of light), Brewster noted that on some reflector plates inclined and placed in a parallel plane to each other, he obtained a peculiar image of the candle he was using as a light source. It was an image of multiple candles approximately grouped around a center point. The following year, in experiments using gold and silver plates, he noted the:

> development [of] complementary colours [resulting in these] multiplied images [being] splendid[ly] tinted, [and] the succession of splendid colours formed a phenomenon which I had no doubt would be considered, by every person who saw it to advantage, as one of the most beautiful in optics.[2]

However:

> notwithstanding the beauty arising from the multiplication of the images, and the additional splendour which was communicated to the picture by the richness of the polarized tints, it was wholly destitute of symmetry; as I was then ignorant of those positions for the eye and the objects, which are absolutely necessary to produce the magical union of parts, and that mathematical symmetry throughout the whole picture, which, independent of all colouring, give to the visions of the Kaleidoscope that peculiar charm which distinguishes them from all artificial creations.[3]

The idea of symmetry is clearly paramount for Brewster's aesthetic conception. It is also evident that he believes that others will find his images as beautiful as he does. In this assumption of universal consent, his comments are similar to Immanuel Kant's view concerning the everyday utterance that such and such a thing *is* beautiful, in the sense such an utterance is a "claim [for] universal validity."[4] In Kant's terms, to say something *is* beautiful is to speak in a way with the belief or expectation that others will agree. Furthermore, it is to suggest that one is not led by a personal interest in the matter. Rather, one is merely commenting on what is obvious or self-evident to all:

> Where any one is conscious that his delight in an object is with him independent of interest, it is inevitable that he should look on the object as one containing a ground of delight for all men. For, since the delight is not based on any inclination of the Subject (or on any other deliberate interest), but the Subject feels himself completely *free* in respect of the liking he accords to the object, he can find as reason for his delight, no personal conditions to which his own subjective self might alone be party. Hence he must regard it as resting on what he may also presuppose in every other person; and therefore he must believe that he has reason for demanding a similar delight from every one. Accordingly he will speak of the beautiful as if beauty were a quality of the object and the judgment logical.[5]

Although Brewster is not as declarative as Kant, his express expectation concerning the beauty of his images is clearly Kantian in the sense of "demanding a similar delight from every one." And although Brewster had an interest in these images by virtue of being the inventor of the kaleidoscope, there is no evidence that he understands himself to be motivated or excessively influenced by this interest. The overall sense is that he is merely remarking on the obvious. But while sharing Kant's view on beauty, Brewster departs most markedly from him on the relation of geometry and beauty. More exactly, he differs on the possibility of there being a science of the beautiful. Kant is quite emphatic that there cannot be a science of the beautiful.[6] By definition, beauty would then be reducible to proof, in the scientific or mathematical use of this term. This would be unacceptable because beauty (and the attending areas of art and skill) would then be reducible to exact or quantifiable knowledge. This would logically lead to the idea that possessing such knowledge and applying it is all that is needed to produce or manufacture beauty.

Brewster, by contrast, sees no difficulty in following this path. As his opening definition attests, he implies that the science (geometry and optics) used in the kaleidoscope's manufacture is the cause of the universal beauty of its generated images. He also implies that beauty is reducible to geometric proof expressed in precise geometric terms. The method through which he advances this science of the beautiful is a detailed program for increasing the precision of the application of geometry in constructing his instrument.

For Kant:

> [the] judgment of taste is not a cognitive judgment, and as beauty is not a property of the object considered on its own account, the rationalism of the principle of taste can never be placed in the fact that the finality in this judgment is regarded in thought as objective. In other words, the judgment is not directly theoretical, nor, therefore, logically, either (no matter if only in a confused estimate) to the perfection of the object, but only *aesthetically* to the harmonizing of its representation in the imagination with the essential principles of judgment generally in the Subject.[7]

For Brewster, the situation is quite the reverse.

Brewster explicitly makes beauty a property of the object. He does this by separating the nature open to ordinary perception from the "symmetrical nature" manufactured within his instrument. Thus, kaleidoscopic perception is geometricized, governed explicitly by geometric principles. These principles, in precise rules of application, function for Brewster as the conditions of possibility for the attainment of beauty quâ visual perfection quâ symmetry. Thus, beauty becomes an actual property of an image by being synonymous with what an image looks like: its visual symmetry. Hence, it follows, that the enemy of beauty is not only the lack of symmetry but also imprecision.[8]

PERFECT SYMMETRY AND ASYMMETRY

In carrying out further experiments, Brewster noted a fact which was to prove decisive in his perfection of the kaleidoscope. Specific liquids, which he was again using to examine polarized light, were enclosed in a trough-like structure formed by cementing side-by-side two sheets of glass. Joined in this way, the sheets formed an acute angle like a V. When he placed his eye at one end of this V-shaped trough, he observed that the cement, pressed through here and there, reflected a pattern on the surfaces of both pieces of glass. This pattern was very much more regular and symmetrical than anything he had witnessed in previous experiments (where he had used objects at a distance to the reflectors). He concluded:

> from the approximation to perfect symmetry which the figure now displayed, compared with the great deviation from symmetry which I had previously observed, it was obvious that the progression from the one effect to the other must take place during the passage of the object from the one position to the other; and it became highly probable, that a position would be found where symmetry was mathematically perfect.[9]

He then set about investigating the decisive features emitting these results, resulting in the following:

> To produce perfectly beautiful and symmetrical forms, three conditions were necessary:
>
> 1. That the reflectors should be placed at an angle, which was an *even* or an *odd* aliquot part of a circle, when the object was regular, and similarly situated with respect to both the mirrors; or the *even* aliquot part of a circle when the object was irregular, and had any position whatever.
>
> 2. That out of an infinite number of positions for the object, both within and without the reflectors, there was *only one* where perfect symmetry could be obtained, namely, when the object was placed in contact with the ends of the reflectors.
>
> 3. That out of an infinite number of positions for the eye, there was *only one* where the symmetry was perfect, as near as possible to the angular point, so that the circular field could be distinctly seen; and that this point was the *only one* out of an infinite number at which the uniformity of the light of the circular field was a maximum, and from which the direct and the reflected images had the same form and the same magnitude, in consequence of being placed at the same distance from the eye.[10]

Hence, obtaining the greatest degree of symmetry requires a precise and fixed geometric relation between the mirrors and the object to be reflected. Moreover, not only are the object and the eye in a fixed relation to each other, but there is only one desired position for each, respectively, in relation to the mirrors. From this notion of a fixed relation and permanently fixed object,[11] and a permanently fixed eye, Brewster's aesthetic eschews variance and movement other than an even or odd ratio of a circle in relation to the angle of the mirrors to each other.

However, the step to bring the simple version of his instrument up to its fullest capacity was yet to be made. He next placed in a cell or cylindrical chamber small objects that could move freely about. This cell was then fitted to one end of the trough formed by the reflector plates. By this move, he introduced a degree of variance into his instrument, but one still well within a traditional geometric horizon: the thin cylinder comprising the cell containing the movable objects.

At this point, Brewster further notes how:

> in this form, the Kaleidoscope could not be considered as a general philosophical instrument of universal application [for any deviation] of the object from the position of symmetry...produced a deviation from beauty and symmetry in the figure.[12]

Hence:

> The next, and by far the most important step of the invention, was to remove this limitation, and to extend indefinitely the use and application of the instrument. This effect was obtained by employing a draw tube, containing a convex lens of such a focal length, that the images of objects, of all magnitudes and at all distances, might be distinctly formed at the end of the reflectors, and introduced into the pictures created by the instrument in the same manner as if they had been reduced in size, and placed in the true position of symmetry. When the Kaleidoscope was brought to this degree of perfection, it was impossible not to perceive that it would prove of the highest service in all the ornamental arts, and would, at the same time, become a popular instrument for the purpose of rational amusement.[13]

What Brewster did here was to construct a hybrid telescope, which, like its older relative, succeeded in bringing the far near. However, while it succeeded in placing, in a reduced form, all the visible mercurial nature within a circular screen, unlike the telescope it was not for the purpose of clarifying what stood at the horizon of ordinary perception. Nor, if we think of the other type of optical screen that reveals, the microscope, was it intended to make visible the teeming world lying below ordinary perceptual horizons. Rather, the kaleidoscope geometricizes perception by producing symmetrical images. It rearranges the perceptual horizon into a symmetric grid. It is, for all practical purposes, a geometricizing telescope, reconstituting along the lines of symmetry the visible nature open to its gaze.

Nevertheless, the issue is not quite so simple. Brewster notes that many objects open to ordinary perception are asymmetrical—or, as he says, irregular.[14] The reflections of such objects in an ordinary mirror, due to being reflected on the posterior surface of the mirror, means that the object and its reflection are always separated in appearance by a space that is equal to the thickness of the reflector or mirror. Furthermore, even if these objects are gathered around a common center "it is impossible by any art to combine them into a symmetrical and beautiful picture [since] the adjacent forms can never join."[15] But since his explanation here is obtuse, let us instead listen to the contemporary scientific theorist, Martin Gardner:

> Asymmetrical solid objects are those that have no plane of symmetry, that can never be made to coincide with their reflections, no matter how they are turned. A simple example is provided by the helix....Try as you will, you will not be able to pass a plane through a helix in such a way that it cuts the helix into two mirror-image halves. Hold a helix up to a mirror. No matter how much you turn it, in the mirror it always "goes the other way...." Two asymmetric figures, each the mirror image of the other, are said to be enantiomorphs. Each is enantiomorphic to the other.[16]

Returning to Brewster: the point he wants to make is that an asymmetrical object, in contradistinction to retaining its asymmetry when reflected in an ordinary mirror, or being merely grouped together with others of a similar appearance, when viewed instead through the kaleidoscope in conjunction with its many reflections therein, is "converted into a...beautiful form," by being combined "with an inverted image of itself."[17] However, for this to happen, the "image must be formed by reflection from the first surface of the mirror, in order that the direct and reflected image may join, and constitute one united whole."[18] Which is to say, whatever object falls under the kaleidoscope's gaze (regardless whether the object is animate, inanimate, or initially asymmetrical) must be made symmetrical. This is achieved by the simple act of duplication and the visual co-joining of the duplicated image with the image of the object being initially reflected. So the object, apart from being simply reflected, is also pictorially joined with its reflected (inverted) double. In this manner, an asymmetrical object is reconstituted into a new symmetrical form by the workings of the kaleidoscope. In short, the resulting composite image is symmetrical regardless of whether the initial object to be reflected is asymmetrical or not:

> The fundamental principle, therefore, of the Kaleidoscope is, that it produces symmetrical and beautiful pictures, by converting simple into compound or beautiful forms, and arranging them, by successive reflections, into one perfect whole.[19]

If we adopt this edict regarding beauty and symmetry, then everything open to the kaleidoscopic gaze *must be made beautiful* by virtue of the geometry governing the reflections and their visual marriage inside this instrument. What better image, then, with which to now visualize Husserl's previous idea that geometry takes for reality what is just a method. It cloaks reality in a garb of ideas. As such, this process of beautification and the kaleidoscope's generated holistic image are eventually reducible to the same mathematical formulation. They are both reducible to the same geometry, permitting the possibility of the instrument in the first place.

In contrast to, say, Castoriadis' always productive but equally reductive plane of abstraction, Brewster's material imaginary bypasses reduction and favors instead beauty and the endless production of symmetry. This is a symmetry that has no need for the Castoriadian idea of differences expressed as unique positions. In the kaleidoscope, differences join hands in symmetry. This joining reduces their difference and emphasizes their sameness. In gathering them all in one position on its reflective screen, it makes a single self-identical object out of them. Thus, Brewster's material version of a planar terrain of an endless production results in collapsing identity, difference, and reducibility into the one process of reflective symmetrization. In this fashion, the kaleidoscope's screen both gathers and hides difference in a totalizing image of beauty, of symmetry.

UNDIFFERENTIATED AND DIFFERENTIATED BEAUTY

Given this totalizing process, does it make any sense to continue to speak of such and such an image as beautiful? Not if we confine ourselves solely to the world offered-up by the kaleidoscope. In this generated world, all is beautiful. The term thus loses it edge, so to speak. It loses its demarcation against some logically necessary corollary: what *is not* beautiful or asymmetrical. To stress this point: Nothing inside the kaleidoscope, no image generated within its reflective borders, can ever be *not* beautiful. This is because all images generated by this apparatus must emerge to the observer's gaze as beautiful quâ symmetrical. They must always be made over as beautiful to the eye of the beholder.

How, then, is beauty to be differentiated if all is beautiful? For the term "beauty" to retain its specificity, the observer must alternate in some manner between the geometrically reconstituted beautiful world imaged in the kaleidoscope and the world open to ordinary perception. At the very least, the observer must retain a recollection of the non-beautiful (asymmetrical) images seen outside the instrument. This has already been granted, by default, by Brewster's earlier comments regarding motivation.

However, if we admit the necessity of an implicit or explicit process of contrast in order to retain coherency in any acknowledgment of beauty, then the question becomes: Given symmetry as inexorably linked to beauty, does this mean that the ordinary world of perception is fundamentally *not* beautiful? Inverting this: What in the asymmetrical ordinary world could ever be beautiful? Very little, it would seem, particularly if we adopt Gardner's view that the universe overall is asymmetrical; his basic view is that things could just as easily have gone the other way, for instance, in relation to handedness.[20] In this regard, Gardner amusingly adds:

> If space is anisotropic, or if asymmetry is built into nature's basic laws, there is no mystery about the absence of antimatter. Physicists, however, have a great fondness for symmetry. Both notions offend their sense of beauty, even though events in the universe seem to go only one way with respect to time.[21]

Is this not the implicit and ultimate promise of the kaleidoscope: to overcome or see beyond an inherently "ugly" world,[22] or not-beautiful (asymmetrical) universe? To be sure, in the sense of making everything symmetrical within its gaze, the kaleidoscope emerges as a physicist's dream. But does Brewster's aesthetic, in being formulated along along the lines of symmetry, not turn back upon itself and bite its own tail? For is his implicit promise to remake the world open to ordinary perception into the image offered up by his scientific aesthetic not forever forestalled because his aesthetic is inexorably *joined* to or contaminated by asymmetry as much as it is marked by symmetry? Which, by extension, is also to say that Brewster is implicitly demanding that what is non-beautiful must also be open to universal consent. Because the objects open to ordinary perception are non-beautiful, due to their asymmetry, they can never by any other means be made symmetrical. Recall that "it is impossible by any art to combine them into a symmetrical and beautiful" image. Only the kaleidoscope can achieve this promise of beauty through symmetricization. Hence, to say that it is ordinarily impossible to make these objects symmetrical is also to imply that their natural asymmetry outside the kaleidoscope is evident to all. Furthermore, the necessary recollection of asymmetry by the observer, so as to maintain coherency in accepting beauty is grounded in symmetry, means that Brewster's science of the beautiful is also, by default, a science of the non-beautiful. This is even more so because his conception of symmetry and asymmetry is a geometric one grounded in the same logical method.

In sum, the geometry informing the kaleidoscope cannot help but construct—or, to make use of a Proclian notion, cannot help but trigger in the mind—the idea of asymmetry at the self-same moment as it constructs an image of symmetry for the eye to behold. Indeed, this play of asymmetry and symmetry as bedfellows in the formation of a privilege granted the latter over the former, only serves as another reminder of how geometry contaminates ceaselessly both the horizon the imaginary wishes to privilege and that to which it sets itself as contrast.

CLOAKING REALITY

The contaminative processes of Brewster's material imaginary get even more complicated by his further desire to take this promise of beautification beyond its simple generation in the screened world of the kaleidoscope. What he has in mind is a comprehensive program of physical reconstitution for beautifying or symmetricalizing a great part of the world open to ordinary perception—specifically, manufactured objects of art and architecture. However, before understanding this program to reconstitute the built world into a likeness of the kaleidoscopic image, we need to understand his wish to overcome the invariance associated with the apparent sameness of the overall patterns generated in the kaleidoscope:

> In all the preceding instruments, the reflecting planes are fixed at an invariable angle...and therefore, though the forms or patterns which they create are literally infinite in number, yet they all have the same character, in so far as they are composed of as many pairs of direct and inverted images as half the number of times that the inclination of the reflectors is contained in 360. It is therefore of the greatest importance, in the application of the Kaleidoscope to the arts, to have it constructed in such a manner, that patterns composed of any number of pairs of direct and inverted images may be created and drawn.[23]

To achieve this end, Brewster goes to some length to detail how the reflector plates may be made movable through a wide range of angles; how more than one center to the generated image may be produced; how to alter the outer form of the image from a circle to other geometric shapes by increasing the number of reflectors; and, finally, how to alter the quantity of light within the pattern by constructing a kaleidoscope from a single prism. Having thus increased the degree of variance of his instrument, he faces one more obstacle: How to free his generated images from their confinement inside the kaleidoscope—how, in other words, to widen the viewing audience beyond a single operator. To overcome this last hurdle, he used a development by one of his contempories who had converted the kaleidoscope into a projector operated by an external light source. Brewster used this conversion as a way to project those kaleidoscopic images, which had initially been derived from directing the apparatus at any object open to ordinary perception (such as trees, statues, flowers, human figures, large pictures, a blazing fire, etc.). These new images were then projected onto a large sheet of paper or some other type of screen, duplicating, for all practical purposes, the camera obscura (except that these projected images were all symmetrical).

Having extended his instrument to capturing most if not all objects open to ordinary perception, coupled with his introduction of a greater variance into the types of overall patterns formed, and expanding the viewing audience beyond a single operator, Brewster is poised to announce what he foresees as the extensive intervention his instrument can make into the physical world of the everyday. He is set to cloak reality in a geometric method. He begins by advancing his instrument as the logical next step in the history of art and natural evolution.

The history of art: any examination of disparate art objects will quickly bring to light, he says, that they all derive their beauty from symmetry. Those exhibiting a perfect development of symmetry are by definition more beautiful. Having almost reduced the entire history and production of art to a developmental process grounded in degrees of symmetry, it comes as no surprise that the kaleidoscope, with its capacity to produce perfect symmetry, emerges as the next stage in this development:

> When we consider the immense variety of professions connected both with the fine and the useful arts, in which the creation of symmetrical ornaments forms a necessary part, we cannot fail to attach a high degree of utility to any instrument by which the operations of the artist may be facilitated and improved.[24]

Nature: nature is also driven by a will to symmetry in the production of its objects. Brewster cites the human figure as having two halves and concludes, "the same symmetry of form presents itself in the shapes of all the various tribes of animated beings."[25] In the case of a plant, due to "the influence of gravitation, or from the slightest breath of wind," the "symmetrical combination of parts has been effected only in its leaves and flowers." Similarly, "when the laws of crystallization are allowed to perform their function uncontrolled, the beautiful geometric forms which they create are marked with the most perfect regularity."[26] However:

> the irregular forms which are the foundation of picturesque beauty constitute a single exception to this general law, and therefore, the Landscape Painter is the only artist who is not professionally led to the study of that species of beauty which arises from the inversion and multiplication of simple forms."[27]

Putting aside this aberration, there would seem, considering the above natural imperative, little room for doubting the historical and, indeed, natural necessity of the kaleidoscope. Against this backdrop, Brewster then draws attention to the social benefits of his instrument by installing it into the very fabric of the industrial society of his time. To accomplish this, he alludes to the social principles governing machinery. Machinery is for:

> abridging manual labour [and it obtains] what is beyond the reach of manual labour...[Machinery also produces an object] in a more perfect state [because it] exhibits a concentration of talent and skill which could not have been obtained by uniting the separate exertions of living agents....[Hence, given that on] this busy island thousands of individuals are wholly occupied with the composition of symmetrical designs [and that there is] scarcely any profession into which these designs do not enter as a necessary part [the] kaleidoscope will assume the character of the highest class of machinery. It will create, in a single hour, what a thousand artists could not invent in the course of a year [while] it works with such unexampled rapidity, it also works with a corresponding beauty and precision.[28]

To add further validation to this exciting prospect, he says artists generally have to wait until their designs are completed before they can say what their final form will take. Because of this restriction, "the art of forming designs...is in a state of extreme imperfection."[29] Now, for Brewster, the desire to know in advance the final form is an inherent desire of all designers, evidenced, he says, by the servile way in which artists and architects of his time copied the Greeks and the Romans. Since the final form of ancient designs is known in advance, to copy them is to assuage the limits imposed upon this inherent desire in the constant search for and development of new designs of which the final form, until now, cannot be known in advance. In contrast, the kaleidoscope:

> when probably constructed, and rightly applied...exhibits the *final* effect of the design...by embodying the very materials out of which the reality is to be produced....[It does this] by exhibiting, instead of lights and shades, the very eminences and depressions which necessarily exist in every design the parts of which lie in various planes.[30]

In this grand vision, it would be easy to overlook Brewster's inventiveness. He collapses the traditional separation between the geometric figures employed in many designs with the actual character or look of the materials to be used in the ultimate construction of the design. He completely erases the need to use geometry's austere figures exclusively as a blueprint one must subsequently extrapolate mentally from in order to "see" what the final thing, in and with all its material peculiarities, may look like when constructed. Conversely, elements of the original material look of the source object are retained in the newly constructed design generated by the kaleidoscope. Hence, no imaginative speculation is called for to envisage what the final constructed form will eventually look like.

In this manner of erasing speculation, the kaleidoscope prises the geometric imaginary from its traditional ethereal seat and places it firmly on a material footing. In one breath, Brewster shifts graphic design elements from austere forms to abundant ones. In so doing, he begins, well ahead of this century, the unseating of reductivism. However, it is only in the field of design that he sets a challenge to reductivism. Conversely, he exhibits no particular explicit desire to forgo reductivism in science. We have already noticed this fact in his reliance on principled geometry to construct his apparatus. Indeed, it is no doubt because of this justifiable reliance on an old geometry that his innovative role in announcing a partial challenge to reductivism has remained obscure since he first proposed it.

IMAGING ABUNDANCE AND GENERATION

If all the above is not sufficiently inventive, Brewster is also able to effect a shift of privilege from austerity to abundance without relinquishing the time-honored plane of the imaginary. The latter is here simply recast as the physical planar reflectors of the kaleidoscope. Therefore, without in any way forsaking traditional geometric principles, by a precise program of application of these principles, he manages to arrange abundant, irregular appearances in such a way, that, for the first time in the history of the geometric imaginary, traditional principles are used instrumentally to generate other than austere, traditional forms. The two-thousand-year-old process of ideation as the exemplary way is here, in this toy, generative of a certain cacography or abundance. Axiomatic geometry producing abundance and not austerity thus permits us to think of this apparatus as a critical empiricism: traditional geometry, in the physical application of this apparatus, critiques itself.

Did this feature contribute to driving the kaleidoscope into the nursery? Where else would such a critical machine be hidden in a century so immersed in the Platonic-Galilean legacy? This was a century that was making such strides toward adding to the successes of traditional geometry by producing such reassuring results and glimpses of the unchanging universal structures seeming to govern nature. Of course, this displacement of the kaleidoscope is not without irony. In many ways, it stood on the cusp or fold in geometry's historical horizon. Around and immediately after the time of the kaleidoscope's invention, the greatest challenge to Greek geometry was taking form: the emergence of other logically consistent axiomatic systems: non-Euclidean geometries.

In weaving all these threads together we may now add to our initial theorization of the kaleidoscope as materializing the geometric imaginary. This miniature hall of reflective screens, gazing out onto the world of ordinary perception, amounts to a theory of creativity. First, generation is grounded in the demonstrative precision that constructs this instrument. Precision forms the groundwork, removing much of the guesswork normally associated with the acquisition and employment of design. Thus, design is here largely mathematized. Vicissitudinous skill is displaced or reconstituted into a program of geometric precision, with novelty being the product of a variance grounded in the infinite combinations of associations. Moreover, regarding the number of images derivable from a single object, Brewster is driven to point out that those ignorant of the true capacity of his instrument have calculated that from "24 pieces of glass, 1391724288887252999425128493402200 [combinations exist;] an operation, the performance of which would take hundreds of thousands of millions of years, even upon the supposition that 20 of them were performed every minute."[31]

For Brewster, this calculation is quite false. It severely underestimates the reality of the kaleidoscope. Rather, "it follows, indeed, from the principles of the Kaleidoscope, that if only one object is used, and if that object is a mathematical line without breadth, the instrument will form an infinite number of figures from this single line."[32] In a word, beauty in its guise of symmetry is endlessly reproducible, and novelty, in its guise of infinite combinations, is endlessly producible and known in advance. The desired end is visible at the beginning.

Second, unlike Proclus, who sought to eschew the generative visible world by grounding geometric generation in an ethereal process of projection onto the screen of the imaginary, which was then viewed by the inner eye of the Soul, Brewster places generation on a material footing, gaining its reflection on physical planar screens open to the outer eye. Also, unlike Serres, who in a reductive manner sought to eschew the wavering outline, the illness of the hand, as a necessary process for the launching of his pure, homeomorphic form, Brewster for his part uses reductivism (principle geometry) to generate an even greater degree of abundance, which ends in a certain critique of the selfsame reductivism.

However, just as Serres' homeomorphic form comes with the demand to overlook cacography, so too does Brewster's schema come with the demand to gather this self-same cacography and reassemble it into an image that seeks to hide certain elements of this irregularity. This hiding takes place in the axis folds of symmetry. Hence, what Brewster reduces is not the graphic form. Rather—now resembling Serres—he erases individuation. Difference at the level of ordinary perceived objects gets lost in his mirrored symmetry. Difference gets reconstituted into an endless set of mirrored differences-of-the-same. The contemporary term capturing the sense (if not the procedure) of this process is "cloning."

What all of this means is that the kaleidoscope explicitly attempts to mirror the abundance of the mercurial, always changing world open to ordinary perception. It quite clearly seeks to see through the traditional veil that geometry used to screen out nature in its past privileging of the ethereal. But it is only a partial "seeing through," because traditional geometry must own up to being the condition of possibility of the kaleidoscope's material construction. It must own up to being the material truth of beauty (reflective symmetry), as well as to being the material possibility of variance (expressed as a set of angular alterations to the kaleidoscope's reflectors, etc.). All of which leads to its owning up to the ceaseless material production of the always "different," yet always the same, novel designs or images.

We now see with greater clarity the meaning of Brewster's opening unequivocal definition of his instrument as that which creates (as much as it exhibits) beautiful forms. Yet, again, this is not quite the whole story.

IDEAL GENERATION

Brewster's theorization of creativity, thought here as a material reflective generation of endlessly variant images, is unable fully to control its own borders. It is unable to close completely its own empirical circle because its material screen is stained with the traces of ideality. Recall that "from the principles of the Kaleidoscope,...if only one object is used, and if that object is a mathematical line without breadth, the instrument will form an infinite number of figures from this single line." Contrary to this curious remark, a line without breadth is not really a material possibility, even for the highly generative screens of the kaleidoscope. The possibility of forming an infinite number of visible figures from such a "line" is therefore nonexistent. This being so, to even imagine such a possibility, and any subsequent generation of an infinite number of figures, emerges as an ideality, or more correctly, a process of ideation. This spectre of ideation is further emphasized:

> This system of endless changes is one of the most extraordinary properties of the Kaleidoscope. With a number of loose objects, it is impossible to reproduce any figure which we have admired. When it is once lost, centuries may elapse before the same combination returns.[33]

Finally, it would not be too much of a stretch to note that Brewster's instrument mirrors parts of Proclus' schema. The kaleidoscope's screens are as much written upon as they write. They are as much written upon by light as they, themselves, in being forever manipulated by the operator, write toward an endless variance of the same. And, in this respect of the empirical being a force in the formation of the imaginary, Brewster successfully adheres to the empirical triggering elements announced by Proclus, at the same time foreshadowing Husserl's much later efforts to bring geometry back to the world from which it first emerged.

METAPHOR

Notwithstanding how the fleeing Platonic clouds seem to be partially regathering on these material screens of the geometric imaginary, threatening perhaps to rush forward and again engulf the empirical, let us continue to use the metaphoric energy of this instrument in our pursuit of the contaminative processes of the geometric imaginary. We can do this by briefly taking cognizance of the common idea of the kaleidoscope as an image for the intense fragmentation of modernity. This cognizance will make apparent that the production of meanings generated by the geometric imaginary will always exceed the tight restraint of geometry's theoretical intentionalities, while appearing to remain forever tied to the resultant images of this intent. Specifically, the ceaseless production of images can generate a slight tremor in the stability of the initial theoretical frame informing the possibility of this instrument.

We will also be considering how the possibility of a seemingly unproblematic correspondence between the ground and its extrapolation, between geometric theory and its applied figures, is intimately dependent upon the control of the truth status ascribed to the figure by the generative theory. Recall that this status in traditional geometry has been consistently understood as the graphic ancilla—a maieutic force, in the Socratic sense of that term—to lawful principles. It follows, therefore, that the possibility of insinuating a tremor into this tradition, by understanding the kaleidoscope's images as announcing more than a mere subordination to geometric theory, must duly excite our attention and interest.

In short, we will be seeking some comprehension of the idea that an image may exceed the prescribed limits of meaning laid down by its own generative ground, while still remaining tied to this ground for its initial and subsequent generation as an image. This generation of excess meaning may be understood as already enfolded *within* the theoretical structure of traditional geometry. To gain a perspective on this enfolded excess, it will first be necessary to examine the common conception of the kaleidoscope as able to image fragmentation and change.

To initiate this examination, we may appeal to a theory of metaphor proposed by Paul Ricoeur:

> [The] metaphor is the rhetorical process by which discourse unleashes the power that certain fictions have to redescribe reality. By linking fiction and redescription in this way, we restore the full depth of meaning to Aristotle's discovery in the *Poetics*, which was that the *poiesis* of language arises out of the connection between *muthos* and *mimesis*.
>
> From this conjunction of fiction and redescription I conclude that the "place" of metaphor, its most intimate and ultimate abode, is neither the name, nor the sentence, nor even discourse, but the copula of the verb *to be*. The metaphorical "is" at once signifies both "is not" and "is like." If this is really so, we are allowed to speak of metaphorical truth, but in an equally "tensive" sense of the word "truth."[34]

Although Ricoeur's argument contains implicitly the conception of a non-metaphoric utterance—linked, no doubt, to a nonfictive reality which, we might assume, he sees as the ground from which his metaphoric process takes its leave toward redescribing reality—it leaves us with the problem of how this grounding reality is to be understood outside metaphoric utterances. But this need not delay us. Rather, what is of interest about his definition is that it draws attention to the two central elements typical of the notion of metaphor in common parlance. The metaphor operates in two registers, those of the *like* and the *unlike*. To achieve this duality, the metaphor needs to be understood as having its own specificity while at the same time referring to a specific referent. Thus, an utterance or an image must in some sense present itself as quintessentially another form of whatever referent it is attempting to allude to or stand for. It must also retain its own specificity to be positioned in the class of visual metaphors associated with the specific referent. It needs to both call up the referent and, at the same time, evoke its own specificity as distinct from any other metaphor claiming to relate to the same referent.

To retain coherence, therefore, the image acting as signifier of some chosen idea cannot, in any fundamental way, be at odds with the idea it alludes to. It also follows that what is at stake here is the coherence of the kaleidoscope's images being understood as images validating the sense of a metaphoric transfer of the conception of fragmentation and change. Putting this another way, does the kaleidoscope's image fulfill the promise of the idea wishing transference? Are fragmentation and change visualized?

IMAGING CHANGE AND FRAGMENTATION

Change: within Ricoeur's demand of *is like* and *is not like*, the metaphoric promise of a distinctive relatedness between the image and the concept is met. The generated image is able to accommodate the two primary comprehensions of the term "change": alteration and process. First, change as alteration: the kaleidoscope's image clearly demonstrates a visual change of state regarding the initial scene or object. Through the act of combining a number of reflections of the initial scene, a composite image is generated that effectively alters the visual extension of the initial scene or object. Moreover, by definition of the very idea of reflection, the original scene is retained as it is multiplied. Thus, the kaleidoscope's image fulfills the metaphoric requirement of *is like* by retaining the visual appearance of the single generating object in the composite folds of the new image. It also fulfills the second theoretical requirement *is not* like by producing a new visual object: the composite image.

Second, change as process: the easily observed characteristic of the kaleidoscope in generating a seemingly infinite number of patterns more than fulfills the specific idea of the ever-changing nature of change. It fulfills the requirement of *is like* by supplying to the viewer a constantly changing pattern *like* that witnessed in ordinary perception. And it fulfills *is not like* by presenting this ever-changing pattern within the confinement of an apparatus whose precise geometric boundaries are seemingly *not like*, for all practical purposes, the "open" world of everyday perception.

Therefore, change in both its registers of alteration and process is adequately imaged by the kaleidoscope within the rigors of Ricoeur's binary definition of the metaphor. The kaleidoscope images an alteration yet also retains what it has altered. It produces an ever-changing image, while this process is forever confined.

However, the metaphoric transfer regarding the process of fragmenting meets a few difficulties. These difficulties are directly associated with the everyday meaning of the term "fragmenting" as an act giving rise to a fragment or a piece of the whole, an incomplete thing or existence. This particular idea of something as being a piece of the whole, or an incompleteness, has traditionally fastened itself onto the comprehension of certain images and assists in ascribing them a particular value. A fragment thought in this way is either valued less because it is not whole or is valued more because it is all that is left of some *original whole* and, therefore, must stand in for this whole. This idea, of a standing-in-for, is predicated on the contradictory claim that an integrity is afforded the fragment that is denied the whole. The whole can be fragmented but not the fragment. Alternatively, if the fragment could be fragmented, then it would have to transform itself suddenly into a whole for the general idea of fragmentation to make any sense. This would amount to claiming that a fragment is that of which it is a fragment. Thus, the inherent logic of this general idea of the fragment, when transported into the visual terrain, demands that the fragment be a type of unframed image—an incompletely framed image, yet also one having a certain corporeal integrity to its borders that is not shared by a whole image. We are asked to perceive, that is, that the fragment-image is somehow an *open closed-ness* in contrast to the *closed closed-ness* of the whole-image.

To unravel this conceptual tangle, it need only be stipulated that while an image can certainly be broken up, what results visually is simply another whole image. Since all images within the kaleidoscope enjoy a visually corporeal integrity, it is absurd to speak of a fragmented image in the above way. This would constitute a claim that an image must have a visually corporeal integrity to be initially defined as such and such an image, yet also *not* enjoy a visually corporeal integrity to be defined as a fragment-image. Conversely, while the composite image formed in the kaleidoscope is obviously made up of many images, they are nonetheless not incomplete pieces of some whole. Rather, they are multiplications of the initial scene open to the view of the kaleidoscope. Hence, what is produced by this multiplication is a visually whole image, not an incomplete piece of some previously imaged whole.

This stipulation permits us to bypass the everyday notion of a fragment as an incomplete part of the whole and to think the term instead at its deeper etymological level: as a *breaking-up*. We must dissociate the idea of an unframed image from the idea of the ability of an image to be broken-up into other images. Of course, this dissociation will undoubtedly lead us to ask: What then has been fragmented? Certainly, not the image of the initial object or scene or their generated reflections. As just noted, an image cannot be a corporeal *incomplete* image, although an image can be broken asunder into a multitude of other whole images. What, then, has been imaged in the kaleidoscope if not incompleteness?

To comprehend the kaleidoscope's generated image in terms of a *breaking-up*, and not of corporeal incompleteness, is to understand that what the kaleidoscope breaks up is visual familiarity or visual expectation. An object or scene, looked at unencumbered by the mediation of the kaleidoscope, would have for all practical purposes a certain look we can assume would be familiar to the viewer. However, when viewed through the kaleidoscope, the image of this self-same object loses its immediate visual familiarity by being visually located in a terrain of its clones. It is, therefore, not the initial object or scene that has been broken up. Rather, what has been broken up is the expectation that a perceptual engagement with a known familiar object will automatically generate a familiar image to perception. The *break* has been inserted perceptually, so to speak, into the register of visual expectation, and not by breaking up the visual completeness of the initial generating object or scene.

In sum, the kaleidoscope's image is metaphorically appropriate by dint of the loss of visual familiarity, and not because the generated image is some incomplete version of the original scene or object. The kaleidoscope simply breaks up the familiar, or to put this another way, it images unfamiliarity. Hence, the kaleidoscope's image *is not like* the concept of fragmentation, if this notion is understood to mean that an image can somehow be corporeally incomplete. Alternatively, the kaleidoscope's image *is like* the concept of fragmentation, if the latter is understood in the sense of its Latin root *to break*. It *breaks up* visual familiarity: it generates unfamiliarity through the process of multiplication, cloning quâ reflection.

With the kaleidoscope capable of imaging change and capable of generating visual unfamiliarity, we are now able to examine the excess of meaning enfolded in the theoretical structure grounding this instrument. To do so, we need briefly to recall the nature of traditional geometry—how appearances (for Plato, in the form of representations) are merely a deficient reality. They hide in their abundance or confusion the eternal principles or laws governing nature (or in Plato's case, that constitute the static, upper echelons of intellection). Axiomatic geometry's promise is simply to overcome this deficiency. In so doing, it generates the condition of a divided graphic terrain. Truthful geometric austere imagery is located on one side, and natural or graphically excessive imagery on the other. Hence, the relationship between austere imagery and geometric principles is itself axiomatic, because of the need in axiomatic geometry to separate eidetic reality from the abundant world of appearances, which is also essential to the truth value it ascribes to its practice.

CRITICAL EMPIRICISM

The matter of excess: In contrast to this promise, the kaleidoscope, while securely grounded in the traditional world of geometric principles applied in a precise fashion, can be manipulated to reorder visible nature into an unfamiliar image. The terrain is visually rearranged according to geometric-optical laws, but not rearranged into traditional geometric forms, although this reordered world is still reducible to that variety of traditional geometric austere forms such as the circle, rectangle, or whatever, comprising the viewing parameters of the ends of the kaleidoscope. Moreover, the patterns generated by this apparatus divide yet further the visual field already divided by traditional geometry, in that the kaleidoscope perceptually reconstitutes nature in such a way that it appears even more chaotic or visually abundant than had been the case within the scheme of traditional geometry—a scheme whose imposition generated the very conception of ideal reality as being the real reality or higher truth of space. But now, far from extracting the truth of space by the employment of principle expressed in austere forms, what emerges through the use of the geometrically principled kaleidoscope is an image of nature that can be said to exaggerate its messiness. The kaleidoscope exaggerates by a series of multiple reflections the apparent visually disordered abundance open to perception even though these multiple or kaleidoscopic images are arranged geometrically, that is, in a symmetrical pattern.

It is in this sense of being able to break up familiar order that the kaleidoscope may be thought of as a form of critical empiricism. It brings into question (through its own material employment) the givenness of the material world open to perception. More precisely, it demonstrates that there is little possibility of an unmediated perception. There is little possibility, solely by an appeal to empirical observation and experiment, of a truth-out-there which is visible to all, mainly because the familiar world opened to ordinary perception is a world already constructed in many ways through the tradition of principled geometry. In fact, this tradition now presents our perception with a wholly different world. It constructs a different (dis)order which is no more or less reducible to this same principled geometry. In short, principled geometry via the kaleidoscope can produce a messy world, just as via other applications it is able to yield an austere one.

In this process, the observer sees within the kaleidoscope the presence (simultaneously and interconnectedly) of both elements of the graphic binary of traditional geometry: the graphically austere and the graphically abundant. The disparate graphic elements of both sides of this divide achieve a marriage in the abundant image of nature being wholly contained within the frame of one of the oldest austerities of geometry: the circle, which constitutes the visual frame of the apparatus. To witness the coextensiveness of these disparate elements is to comprehend the generated image and its austere containment as an ordered disarray—visible nature reordered by geometry, geometrically jumbled up and encased in, encircled by, an austere geometric form: a symmetrical mess. Through this generation of a certain chaos inside a traditional form, geometry both disavows and maintains its tradition.

If we allow that austere form is now cojoined with abundant imagery, with the latter gaining visual center stage, then the kaleidoscope can be thought of as an apparatus that goes a long way toward reversing the binary structure of traditional geometry. Abundant visible nature, when viewed through this apparatus, now takes up visual residency within rather than without the austere circle and demonstrates that lawful principle contains the possibility of a differently ordered abundance. Indeed, it is abundant appearances that now attract attention and become, as it were, more real than the austere framework. Traditional principled geometry acts in a maieutic role to assist in the birth of an ordered disarray rather than being an exemplary force for the attainment or comprehension of ideal reality. In this fashion, the vision afforded by the kaleidoscope heralds contemporary geometric interest in the lawfulness or order in messiness.

Thus, the austerely shaped screen of this materialized geometric imaginary is stained through and through. It is stained all across its surface with the previously underprivileged, displaced, abundant world of visible nature. From such a contaminative configuration, it is easy, therefore, to imagine that geometry completes a full circle and generates the very thing it initially sought to displace so as to establish the primacy of principle. It is in this sense of the unwanted return of the displaced other that traditional geometry via this instrument exceeds its own theoretical ground. Indeed, can we not see clearly here, within this instrument, the very limits or edges of traditional geometry, its ability to make a mess of things? Can we not hear the announcement, in this generated image, of the limits of traditional geometry's eidetic reality as speaking the truth of space over perception? Through the employment of this instrument, may not the observer gain the clear perception that lawful principle is in some sort of historic twilight and has become simply a generative force for visual unfamiliarity?

HISTORICIZATION OF MESSINESS

The vision afforded by this instrument challenges the contemporary idea that geometry's interest in messiness (fractal and chaotic theories) is strictly contemporary. Does not the kaleidoscope's image, in its ability to announce the closure of lawfulness in geometry and its rebirth—the shift from law-as-principle to law-as-process—does this image not stand as some haunting presence insufficiently captured by geometry's ongoing historicization of its own practice? How, considering the persistent presence of this nursery plaything, will the discourse of geometry historically locate with assured security the origin of its interest in lawful messiness? How will it locate its current interest in change and its processes? Will this interest be historicized in fractals, or in this philosophical toy? Will it be in the recent discovery of "nature's own geometry?"[35] Or will it be in the laughter of some child gazing at the visual break put into the long unfolding of geometry's promise of truth through the precise application of its principles?

When the history of geometry is written, will we hear the distant laughter of ideal geometry as it acknowledges itself as the author of the child's delight when he or she gazes through this plaything and sees the ever-changing fragments of the old promise of an austere, reductivist's truth to space? Conversely, within the contemporary logic of nature having its own geometry, will not laughter and the chaos of toys themselves be also open to geometricization? Finally, will this instrument constantly serve to remind us that geometry's interest in the irregular has always been part of its horizon, merely awaiting colonization, awaiting the development of a geometry of irregularity?

LAUGHTER IN THE RUINS

Speaking of laughter, we might see how appropriate is Brewster's naming his apparatus a philosophical toy. What image in the history of geometry could be more hysterically philosophical than that generated by the kaleidoscope? Recall: An empirical image grounded in the ideal that questions its own ground at the very same moment as it toys with the possibility of a fallback into this self-same ground by wrapping itself in an austere circle. A finite instrument contaminated by the infinite (remember Brewster's remarks regarding the number of images capable of generation). Infinity generated within a circle! Moreover, this generated infinity is ever-changing, yet in a sense remains always the same.

Little wonder then, that this scientific apparatus should find its way into a nursery. Where else than in a nursery, where the conception of a universal truth to space is perhaps yet to get a firm hold, could we imagine the existence of such an instrument creating such delight? And in this delight, might not it be imagined that one faintly hears, while gazing into this miniature hall of mirrored ruins, the distant laughter of philosophy as it muses over how, in this toy, the ideal in principle is good for making things, and the empirical in application for unmaking them? Indeed, what instrument could have toyed with philosophy more—or been banished so quickly?

Is it perhaps also not laughingly appropriate that the kaleidoscope has now disappeared from the nursery, overwhelmed by and hidden beneath the chaotic swaddling and educational electronic abundance commercially constituted by Toys R Us and its manifold imitators? Is it not the kaleidoscope that heralded the postmodern culture of adult children in a world of fragmented tourist symmetry, where Holiday Inns are available in every city throughout the world? And is not a chuckle appropriate when contemplating the kaleidoscope of sports bars, and their ubiquitous televisions twirling night and day, universalizing the explosions of beer ads? As for the kaleidoscopic blooming of Gap clothing stores, and terrorist movies, that all seem the same in their differences, are they not precisely the ultimate comedic irony of Brewster's invention by being housed and framed in culturally transcending environments of safe, airconditioned fax availability—halls of mirrors by other names?

CODA

However, putting aside the revolutions hiding in nurseries, it is important to reemphasize how this beguiling instrument marks a start in scientific efforts to pry the geometric imaginary out of its ethereal and reductive abode. As an instrument of thought, to be sure, it is quite rudimentary. Nevertheless, in all its simplicity, it still manages to gather the key elements of a tradition of thinking the geometric imaginary handed down to us from the time of Proclus, at the same time as it reconstitutes them on material reflective screens. The traditional form of these elements are:

1. The geometric imaginary is a mental screen reflecting the ethereal.
2. This screen interposes itself between two separate states or realities (static truth and mercurial opinion).
3. The images or figures inhabiting this traditional screen initially come from one or the other of these separate realities.
4. The geometric imaginary is self-generative.
5. The images inhabiting the geometric imaginary are endlessly repeatable, but self-same (more or less).
6. These images are more pure than ordinary perceptual and physical images; hence, this screen and its images emerge as an approximation of, or an exemplary way to glimpse, a utopian or ideal horizon.
7. The geometric imaginary constitutes a theory of creativity grounded in sameness.
8. Finally, at the edge or in the shadow of this tradition, this screen is constantly being contaminated or stained by the impurity to which it sets itself in contrast or from which it separates itself.

Brewster's reconstituted version of the traditional elements, founded on his invention of the kaleidoscope, is:

1. The geometric imaginary is a material screen which by the process of reflection reconstitutes the mercurial world open to perception.
2. The screen interposes itself between the eye and visible nature (static condition and mercurial possibility).
3. Nature is the sole source of the images on the screen.
4. Generation occurs manually.
5. The screened images are endlessly different but their overall composite pattern is endlessly repeatable.
6. The images on the screen are universally beautiful and always symmetrical in contrast to the asymmetrical objects visible in nature; hence, a utopia of symmetrical design is proffered as realizable.
7. The screen is a theory of creativity grounded in novelty.
8. Finally, the material screen of abundance is stained—first, by the fact that the possibility of its construction is grounded in principled (ideal) geometry; second, by a necessary recollection of asymmetry as a condition for the viewer to accede to Brewster's privileging of symmetry; and third, by an ideality that inhabits it as the possibility of an infinite production of images.

This early effort to grant a materiality to the geometric imaginary was nevertheless rudimentary in at least one respect. Where Brewster's imaginary differs most from the traditional elements, apart from its materiality and production of abundance, is in the area of generation. His screen is not entirely self-generative in the sense of Proclus' model (allowing that Proclus' screen may, but need not, be triggered by an outside, or empirical agency). Brewster's generation requires physical force, which stands outside the parameters of his screen. This is, of course, the operator who twirls the kaleidoscope to generate change in the patterns being produced. With this limit in mind, what is needed to complete the full materialization of the geometric imaginary is a screen which is, for all practical purposes, self-generative. The obvious candidate is the computer and its digitally programmed screen.

NOTES

1. David Brewster, *Treatise on the Kaleidoscope* (London: Archibald Constable, 1819), p. 1.
2. Brewster, *Treatise*, p. 2.
3. Brewster, *Treatise*, p. 3.
4. Stephen Körner, *Kant* (Harmondsworth, Middlesex, UK: Penguin, 1955), p. 183. Brewster was a member of the common sense philosophers who were strongly influenced by Kantian philosophy.
5. Immanuel Kant, *The Critique of Judgement*, trans. J.C. Meredith (Oxford, UK: Clarendon Press, 1986), p. 51.
6. Kant, *Critique*, p. 165.
7. Kant, *Critique*, p. 216.
8. Brewster, *Treatise*, p. 4.
9. Brewster, *Treatise*, p. 4.
10. Brewster, *Treatise*, pp. 4-5.
11. Brewster, *Treatise*, p. 5.
12. Brewster, *Treatise*, p. 6.
13. Brewster, *Treatise*, p. 6.
14. Brewster, *Treatise*, pp. 15-16.
15. Brewster, *Treatise*, p. 16.
16. Martin Gardner, *The Ambidextrous Universe*, 2nd ed. (London: Penguin, 1979), pp. 19-20.
17. Brewster, *Treatise*, p. 16.
18. Brewster, *Treatise*, op. cit.
19. Brewster, *Treatise*, p. 17.
20. Gardner, *Ambidextrous Universe*, pp. 221-231.
21. Gardner, *Ambidextrous Universe*, p. 230.
22. Brewster, *Treatise*, p. 64
23. Brewster, *Treatise*, p. 79.
24. Brewster, *Treatise*, pp. 113-114.
25. Brewster, *Treatise*, p. 114.
26. Brewster, *Treatise*, op. cit.
27. Brewster, *Treatise*, op. cit.
28. Brewster, *Treatise*, pp. 115-116.
29. Brewster, *Treatise*, p. 116.
30. Brewster, *Treatise*, pp. 116-117.
31. Brewster, *Treatise*, p. 111.
32. Brewster, *Treatise*, p. 112.

33. Brewster, *Treatise*, p. 113.

34. Paul Ricoeur, *The Rule of Metaphor: Multi-disciplinary Studies of the Creation of Meaning in Language*, trans. R. Cernzy (London: Routledge and Kegan Paul, 1978), p. 7.

35. Benoit Mandelbrot, *The Fractal Geometry of Nature* (New York: Freeman, 1983), pp. 1-33 and C1-16 between pp. 276-277.

Illustration 11. Front plate from William Blake's *Europe: A Prophecy* (1747) conflated with Benoit Mandelbrot's *The Fractal Geometry of Nature*, updated and enlarged (New York: W.H. Freeman, 1983), especially pp. 49 and C1. A sketch for Blake's design in his *Notebook* is titled: "Who shall bind the Infinite." (Drawing by James Frazier.)

Chapter 7

BENOIT MANDELBROT'S GEOMETRY OF NATURE

CRITIQUE OF AUSTERITY

The idea that the world of the experiential should be the first and foremost concern of geometry is the central theoretical claim advanced by the geometer, Benoit Mandelbrot. Highlighting this claim, he writes:

> Why is geometry often described as "cold" and "dry"? One reason lies in its inability to describe the shape of a cloud, a mountain, a coastline, or a tree. Clouds are not spheres, mountains are not cones, coastlines are not circles, and bark is not smooth, nor does light travel in a straight line.
>
> More generally, I claim that many patterns of Nature are so irregular and fragmented, that, compared with *Euclid*—a term used in this work to denote all of standard geometry—Nature exhibits not simply a higher degree but an altogether different level of complexity. The number of distinct scales of length of natural patterns is for all practical purposes infinite.
>
> The existence of these patterns challenges us to study these forms that Euclid leaves aside as being "formless," to investigate the morphology of the "amorphous." Mathematicians have disdained this challenge, however, and have increasingly chosen to flee from nature by devising theories unrelated to anything we can see or feel.[1]

Mandelbrot adopts here the standard view of traditional geometry: the view it has of itself. Namely, traditional geometry is taken as the voice of austerity. It is solely the speaker of idealities and is unrelated to visible nature in all its mercuriality. Traditional geometry has nothing to say about the irregular, the cacographic. It is stuck on the static. In adopting this view, Mandelbrot turns the time-honored strength of traditional geometry, the unchanging character of its austere shapes, back upon itself as a limitation. Static, unchanging figures or shapes have nothing to say about change. A geometry of idealities has nothing to say about perceived, mercurial reality.

Having thus launched his critique on the question of shapes and limitations, he says that, in contrast, some fractal sets "are so oddly shaped there are no good terms for them in either the sciences or the arts."[2] Indeed:

> [S]cientists will (I am sure) be surprised and delighted to find that not a few shapes they had to call *grainy, hydralike, in between, pimply, pocky, ramified, seaweedy, strange, tangled, torturous, wiggly, wispy, wrinkled*, and the like, can henceforth be approached in [a] rigorous and vigorous quantitative fashion.
>
> Mathematicians will (I hope) be surprised and delighted to find that sets thus far reputed exceptional...should in a sense be the rule, that constructions deemed pathological should evolve naturally from very concrete problems, and that the study of Nature should help solve old problems and yield so many new ones.[3]

What is so important about Mandelbrot's view is that geometry is no longer reducible to a handful of austere figures. The figural complexity inherent in anything described as "grainy" or "wiggly" immediately suggests that in constructing and coming to terms with these forms, geometers will no longer be required to abstract-out or imagine morphological forms standing over and above the irregularity of physically rendered shapes. They will no longer be obliged to see a universal, ever repeatable austere regularity behind irregular graphic figures. Indeed, it is precisely these latter shapes in all their wiggliness and graininess that are now to be thought of as the province of geometry.

Moreover, since most if not all of these new geometrical figures are found in the "study of nature," they must be open to ordinary perception and thus, by implication, automatically and immediately intersubjective and intercommunicable. The complexity of fractal shapes poses no real communicative obstacle precisely because they are, in a sense, already familiar.[4]

GEOMETRY OF THE AESTHETIC

This sense of the shared and immediately intersubjective communicative character of fractals (universal consent) is nowhere more apparent than in the following citations:

> This essay brings together a number of analyses in diverse sciences, and it promotes a new mathematical and philosophical synthesis....Furthermore, it reveals a totally new world of plastic beauty.[5]
>
> [M]y previous Essays stressed relentlessly the fact that the fractal approach is both effective and "natural." Not only should it not be resisted, but one ought to wonder how one could have gone so long without it.[6]
>
> In addition, fractal geometry reveals that some of the most austerely formal chapters of mathematics had a hidden face: a world of pure plastic beauty unsuspected till now.[7]

For Mandelbrot, "fractals tend to be astoundingly beautiful."[8] Moreover, "In particular, in the theory of fractals 'to see *is* to believe.'"[9] Finally:

> The "fractal new geometric art" shows surprising kinship to Grand Masters' paintings or Beaux Arts architecture. An obvious reason is that classical visual arts, like fractals, involve very many scales of length and favor self-similarity. For all these reasons, and also because it came in through an effort to imitate Nature in order to guess its laws, it may well be that fractal art is readily accepted because it is not truly unfamiliar.[10]

Like Brewster, Mandelbrot's aesthetic exists within the Kantian tradition. Indeed, Mandelbrot's famous declaration, "to see is to believe," leaves little room to doubt he is making the assumption of an immediate universal consent with regard to the naturalness of his fractals. His whole program is precisely this: a geometry of the aesthetic which, he feels, constitutes a radical departure from traditional systems of measurement grounded solely in an ideal, static, irreal phenomenon. Of course, what is disavowed in this schema is any role for a critical reading of fractal images—a role that would clearly pose some danger for geometry, and perhaps even science in general, since it would create the possibility for the various discourses of visual studies to intervene in geometry at the level of its theory formations. Were the images of fractal geometry critiqued in such a way as to call into question their truth claims, then it is difficult to see how fractal geometry could maintain that its images mirror the "fractal face of nature." Were such a critique to succeed, fractal geometry might well find itself forced to withdraw its most radical claim.

But such a withdrawal would raise the possibility of comprehending fractal geometry as merely another theory of representation among many. This result would severely damage the one-to-one correspondence with nature central to Mandelbrot's challenge to traditional geometry. To what degree this impasse, brought on by Mandelbrot's preference for declarative statements, will ultimately emerge as a serious flaw pointing to the failure of contemporary geometry to engage the long history of the humanities' theorization of the experiential, is a matter beyond the present. But what is germane is that geometry's interest in speaking the truth of the world has not abated. Indeed, it has intensified. Indeed, it has materialized this interest.

MONSTERS

Mandelbrot duly acknowledges that some irregular (proto-fractal) shapes have previously emerged in the domain of traditional geometry. He refers several times throughout his book to the fact that these "monsters" (a term applied by Poincaré) emerged in nineteenth-century mathematics, but in an unwanted fashion. Although nineteenth-century mathematicians may have thought that these monsters were pathological forms, unwanted orphans, Mandelbrot has something altogether more positive in mind. These early monsters, now augmented by his fractals, form the basis of a new science. A science of *the nature and causes* of those shapes which traditional geometry initially displaced precisely because of their pathology. This new *pathognomonic science* is one that can approach these grainy and wiggly figures in a "rigorous and vigorous quantitative fashion."

In being occupied with the wiggly and the grainy, this new science will implicitly remedy, restore to geometry, the immediate world of the senses. What is being brought back, restored to prominence, finally brought in under the mantle of the quantitative, is precisely most (if not all) of the figural irregular, the qualitative. In a word, fractal geometry will remedy (in the other sense of that term, as fixing-up) the quantitative gap left open by Galileo's mathematization of nature in its inability to completely mathematize the plenum.

FRACTAL MANTLE

It is important to stress the intellectual depth and theoretical audaciousness of what Mandelbrot is proposing. If qualities can be thought of as properties of objects and are necessarily tied to shapes, then by definition, if *natural* shapes can be manufactured, then so too will all of their accompanying qualities fall directly under the (fractal) mathematical mantle. The ancient idea of correspondence expressed as mathematical indices between changes in shape and plena will be revamped. Shape and plena will be a direct expression of, and produced by, the same (fractal) mathematics.

In sum, Mandelbrot is taking irregularity, the cacographical, the qualitative (primarily the aesthetic) and making them the explicit subject matter of geometry. What, then, would such a science of graininess and wiggliness look like? What image does Mandelbrot offer of this science that seeks to restore or explicitly write the irregular back into the equations of geometry? What is the character of this *pathological geometry*?

The most significant image offered is philosophically provocative, even startling. What Mandelbrot seeks to displace from the discourse of geometry is nothing more or less than the idea that the transcendent truth of space is adequately expressed as an ideality. More precisely, he replaces the traditional abstract imagination with what amounts to an empirical imagination. This displacement of abstraction and all its contents is enacted by a persistent privileging of the perceptually obvious. This displacement is encapsulated in the following succinct phrase: "In brief, I have confirmed Blaise Pascal's observation that imagination tires before Nature (*'L'imagination se lassera plutôt de concevoir que la nature de fournir'*)."[11]

Such an observation allows Mandelbrot to lay the ground for a new type of geometric imagination. He begins to map the character of this new imaginary with a few qualifications concerning the role of the image in his schema:

> However, showing pretty pictures is not the main purpose in this Essay; they are an essential tool, but only a tool. One must also recognize that any attempt to illustrate geometry involves a basic fallacy.[12]

The tension between the use of the terms "essential" and "fallacy," coupled to the fact that the term "geometry" in this context implies a mental construct somehow more truthful than any graphic representation thereof, raises an obvious question. In what manner are we to think of fractal images as an essential tool while accommodating his other idea that imaging geometry constitutes a basic fallacy? How, despite this fallacy, or perhaps precisely because of it, is Mandelbrot to implement the task of affording fractal images a higher status than the austere diagrams of traditional geometry?

HAND DRAWINGS

Mandelbrot initially forges his critique by noting the gap between traditional geometric figures and perceived nature. He closes this gap by displacing the abstract geometric imaginary and introducing his empirical geometric imaginary, the images of which are identical with those open to perception. He also protects this closure by assigning universal consent both to the naturalness and the beauty of these images of identity.

Recall that his overall task is to map geometrically what we "see" and "feel." He desires to geometricize the mercurial world open to the Subject, the world the Subject experiences in its "inner" world of feelings and in its "outer" world of seeing. He is directing the attention of geometry toward the processes that issue or constantly produce the world to the Subject as this or that world.

Given this approach and the substance of his critique, Mandelbrot is poised to take the next and crucial step: affording his fractal images a higher status than the austere figures of traditional geometry:

> Thus the typical fractal curve can be evoked satisfactorily by elementary strokes in large but finite number....The larger the number of strokes and the greater the accuracy of the process, the more useful the representation, because fractal concepts refer to the mutual placement of strokes in space, and it is vital in illustrating them to keep to precise scale. Hand drawings would be prohibitive, but computer graphics serves beautifully....Also, I am fortunate in having access to a device that produces camera ready illustrations....Graphics is wonderful for matching models with reality....A formula can relate to only a small aspect of the relationship between model and reality, while the eye has enormous powers of integration and discrimination.[13]

The point of the exercise is to render fractal curves in such a way that the detail on each successive level or scale is captured exactly. Scaling, in particular the self-similarity of complexity evident on successive levels of scale (successive scales exhibit the same degree of complexity), is precisely the unique feature to which Mandelbrot has alerted the discourse of geometry and which it is desirous of rendering accurately. Therefore, with regard to his fundamental demand regarding detail at successive levels of scale, drawing by hand is simply unable to cope. Moreover, since fractal curves are complex in appearance, in contrast to those of traditional geometry, then a form of rendering is necessary that will do adequate justice to this complexity. It follows that the temporal constraints imposed on manual drawing, even when assisted by such traditional instruments as the rule and compass, further undermine the power of the traditional way of rendering the geometric imaginary.

The very simplicity of the move here, grounded as it is in the indisputable physical limitations of manual drawing, belies the theoretical complexity enfolded in his displacement of the traditional form of rendering geometric concepts. Manual drawings are not being displaced because of their irregular features. Indeed, explicitly writing irregularity or complexity back into the discourse of geometry is precisely the feature that constitutes one of Mandelbrot's innovative moves. Rather, manual drawing is impugned because of its inability to be sufficiently irregular. Manual drawings are simply not irregular enough, and hence cannot adequately capture the complexity or irregularity of fractal curves. In this one clever stroke, Mandelbrot positively retains cacography while displacing geometry's traditional figures and the technology of their rendering. This displacement at the very same time affords him the opportunity to introduce a new way to do geometric rendering: the computer.

THE ELECTRONIC SCREEN

In the process of offering the computer as a desirable, even a necessary aid in the rendering of fractals, Mandelbrot sweeps up computer operators in the wake of this announcement and installs them as ancillary or maieutic workers who (as he duly acknowledges) importantly lend their creative and technical skills in bringing to the fore the images of this new geometry.

However, it must be stressed that the emphasis is on the image not the operator. In other words, Mandelbrot also shifts the emphasis from the actor to the enacted—relocates the ancilla, as it were. It is the operators who emerge now as the ancillas, handmaidens to the attainment of—or more properly, the computing of—the (fractal) truth of space. The whole purpose of this contemporary (now non-Socratic) midwifery is to render truthfully and exactly on computer screens an image of the world and its processes open to the gaze of ordinary perception. These programmers are in the service of perceptual reality, in contrast to the images of traditional geometry, which were in the service of an ideal reality.

In Mandelbrot's schema, then, it is not the graphic image that assists in the birth of the truth or acts the role of ancilla. Instead, the computer image is the truth; it mirrors or looks exactly like what we already see. The material image is not a thrice-removed truth, as was the case with Proclus' schema. It is, instead, now a first order truth. This is a first-order status, which Mandelbrot achieves by collapsing the Proclian third-order truth (perception) with Proclus' second-order truth (the geometrical screen of the imagination), leading to a reversal of the structure of the traditional schema in a way that places at the top what for Mandelbrot we naturally see, cojoined with what from all these natural sights can be generated on a computer screen. In turn, what is relegated to the tail end of reality, at the lowest rung of the contemporary hierarchy of geometric reality, is abstraction and all its attending purity, regularity, staticness, and so forth. But the fractal image generated by the computer can only be a first-order truth if we accept it as unproblematically rendering what we see (and its complexity) in ordinary perception.

However, the relation between what we "see and feel" (Mandelbrot's phrase) and the "complexity of what we see and feel" is anything but unproblematic—although they often appear to be the same for Mandelbrot. The rendered image on a screen cannot completely mirror the actual materiality of any perceived fractal form in nature. From this perspective, the computer-generated image can never speak the truth of a natural fractal if, as in Mandelbrot's logic, truth is an initially experienced truth and, therefore, necessarily attended by the materiality that we may assume is part and parcel of the world open to ordinary perception. That is, the computer-generated fractal can never be an exact replica of what we "see and feel" to be nature's own geometry.

The fractal image as a mirror for natural truth is therefore something of a chimera—or, more exactly, a mirror whose light shades as much as it illuminates. Thus, in much of Mandelbrot's writing, the gap simultaneously grows and shrinks between understanding the computer-generated fractal image as a first-order truth (truthfully mirroring nature's fractal face, truthfully imaging what we "see and feel"), and understanding it at another level as a second-order truth due to its lack of being able to replicate in all their materiality the objects it is mirroring. In short, to see the computer image as a first-order truth faithfully mirroring what we naturally "see and feel" is dependent upon our suspending elements of what we naturally "see and feel."

In order to mirror the fractal face of nature and to privilege the mirroring technology over the abstract or reductive technology of traditional geometry, Mandelbrot must employ the very method that his fractal geometry is desirous of critiquing, namely, traditional geometry's suspension of elements of what we "see and feel" in order to make possible the seeing in the mind's eye (instead of on the computer screen) of the most true realities. In Mandelbrot's case, we have to not see, not take account of, suspend from our thinking, the materiality of the perceived thickness of natural fractals in order to see the computer image as a truthful image of a fractal in the world. In this precise sense, we are blinded by the illumination of the computer screen. "To see is to believe" may yet turn out to be a demand grounded in suspending quite a lot of what is believed to have been seen or felt (Serres' radical forgetting is echoed here).

Notwithstanding this problem of suspension, Mandelbrot continues to forge a reenchantment of the image by installing computer graphics as the intermediary between geometric concept and reality. In this installation, he manifests an inverted spirit of Proclus' intermediate mathematical projection. Although Mandelbrot's computer screen acts as an interface between two realities (his intuitively obvious geometrical concepts and the world open to ordinary perception), it is unlike Proclus' geometrical plane of the imagination in the first instance because it is physically external to or outside the Subject. Moreover, it is a material screen that mirrors mental (fractal) concepts and the world open to ordinary perception. As it "matches" one with the other, it in a sense collapses both into itself.

Mandelbrot's empirical screen pictures what we think and what we see in the field of perception. In this mirroring, both are reduced to the computer screen. They meet or interface (a telling phrase) at the surface of the computer. This is brought home forcibly—and here the graphic is fully reenchanted—when we recall that "a formula can relate to only a small aspect of the relationship between model and reality while the eye has enormous powers of integration and discrimination."[14] What this suggests is that computer graphics resolve any tension between mental models and perceptual reality in favor of the latter. Computer graphics arbitrates toward what we "see," toward the visual—that is, toward itself.

In light of this arbitration toward itself, it is perhaps not too extreme to imagine that the computer marks the possibility of producing, within its seemingly endless programmable generative capacity, what will count as reality—or, more exactly, what will count as the real by virtue of what it can geometricize in the imagery of fractal figures. Indeed, we may choose to think of this electronic screen as a programmable external imagination, with all the creative capacity of the traditional imagination now reconstituted into the almost endless generative capacity of the computer in a manner that gives form to the true reality. The computer becomes a screen that speaks (shows) the truth of space as now an empirical quâ perceptual truth. This may yet turn out to be the same as saying that fractal imagery within Mandelbrot's schema turns contemporary geometry into an explicit scientifically subscribed subjectivity quâ experience quâ perception.

To stress this latter point: Mandelbrot speaks of the "eye" as almost a disembodied entity with "enormous powers." He writes as if it were an independent organ with attributes and capacities shared by all subjects, yet with habits disassociated from any connection with the mental activity of abstraction in such subjects. He also suggests that computer graphics resolve or collapse difference toward the visual. Coupled thus with the fact that computer graphics are to be understood as mirroring what the eye sees, the computer screen emerges as standing in a certain equivalence to Mandelbrot's eye. In the employment of computer graphics, the eye is made manifest at the level of the material. The eye of the Subject sees, in a manner of speaking, from without the Subject. Insofar as the eye of the computer looks back at each and every Subject and (to borrow a Proclian term) projects onto the Subject's eye the truth of nature quâ perception, then Mandelbrot's material geometric imagination (or computer screen-graphics) participates as much in the formulation of what the Subject sees as it mirrors what is seen. Thus, in Mandelbrot's reversal of geometry's binary, he unintentionally returns to Proclus' projection, although from Mandelbrot's perspective the truth of space is projected up from an empirical reality and not down from an ideal one.

The geometric screen of the imagination is therefore no longer a timeless internal space of austere abstraction through which the Platonic mind's eye can glimpse the essentialities or principles that govern Being. Rather, it has reconstituted itself into a temporal external space that, for Mandelbrot, mirrors the fractal processes from which and out of which Being emerges and wends its way. One might almost say that the Mandelbrotlian Subject looks for Being on the screen. In this fashion, the Subject's own nature and lived-temporality is brought in under the mantle of the mathematical fractal image—becomes an expression of "nature's own geometry." Or, to borrow freely from Husserl, one could say that experience itself becomes a fractal manifold.

SELF-GENERATION

The only thing left, then, to complete the materialization of the traditional geometric imaginary is to grant it a capacity for self-generation. This presents few if any problems, since the virtually endless generative capacities of the computer mark it out from all other scientific instruments. Equally, its generation is determined in advance because of the logic of its system. In this sense of computer generation being a determining-in-advance, Mandelbrot's material imaginary faithfully (if unintentionally) maintains all the elements (save overt ethereality) that inform the inherited or handed-down tradition of the geometric imaginary. However, in light of this tradition, how is geometry's electronic imaginary to maintain itself as still the exemplary way, the royal road to truth?

THE ROYAL ROAD

Clearly, it cannot be exemplary by virtue of reflecting the lasting character of universals, if by universals we mean, in the old sense, unchanging, static idealities—out-of-sight truths that are invariant. Fractal geometry does not look through the veil of nature and see its inner invariance. Quite the contrary. It is precisely variance that it wishes to model or mirror. This is a geometry seeking to image change, in contrast to two millennia of imaging samenesses.

Moreover, what is repeated on this electronic imaginary is not Castoriadis' miracle of the difference of the identical. Rather, it is the miracle of the production of differences from the identical (similar degrees of complexity). In this manner, the universal is refigured from being the law-as-principle to the law-as-process. The truth of the universe is no longer to be thought an out-of-sight reality which geometry permits us the opportunity to glimpse. *The truth is precisely what we already see.* The age-old dream of geometers and philosophers of glimpsing nature's truth turns out to be a truth already visible, in plain sight (rather than sighted on an ethereal plane). The new task of geometry is to formalize this in-sight truth. The royal road to truth is the road itself.

THE DIVINE SCREEN

What, then, stains this electronic imaginary that images gross impurity? What trace marks its surface? Although the following is a rare piece of geometric humor, its significance for inclusion is not simply one of light relief. Rather, it is a paraph, or stroke, which Mandelbrot draws under his fractal signature of nature—a stroke or flourish that images the stain or trace of what is perhaps the meta-ideality of all idealities: namely, God.

In his remarkable book, Mandelbrot makes use of an image that is the frontispiece of the Bible Moralisée (circa 1220-1250). The image is of God drawing, with a pair of ancient compasses, what presumably is a depiction of the cosmos. The original legend, Mandelbrot informs us, reads, "*ICI CRIE DEX CIEL ET TERRE SOLEIL ET LUNE ET TOZ ELEMENZ*" (Here creates god sky and earth sun and moon and all elements). In employing this image, Mandelbrot changes the legend to: "HERE GOD CREATES CIRCLES, WAVES, AND FRACTALS."[15]

Having with considerable ingenuity laid a serious challenge at the door of idealized and axiomatic geometry, Mandelbrot cannot then hold his challenge free from the contaminative presence of what he has sought initially to displace. Here is an image that has ideality as the initial generator of fractal geometry. First, a supreme ideality (God) generates geometric idealities (waves and circles). Second, God generates geometric irregularities (imaged as wiggly shapes comprising part of the cosmos's matter). And all of this is contained inside one of the oldest planar geometric forms in history: the traditional circle enclosing the cosmos God is drawing (an austere enclosure, prefigured, recall, by the kaleidoscope—and may we ever think anything but an austere figure will suffice to image totality?). And to cap this return to the past history of geometry, God does all this with the humble compass! In sum, this divine image turns Mandelbrot's critique back upon itself by making an ideality (God) the generator of his fractals.

STAINS ON THE SCREEN

In now returning to the ideal, in tracing our own circle of containment, we remain poised or suspended between two horizons—two screens of the geometric imaginary—both of which are stained by each other, both of which bear the footprints of the other dancer. The *irregular* horizon says:

> We perceive three different kinds of form in this newly created world: circles, waves, and "wiggles." The studies of circles and waves benefited from colossal investments of effort by man, and they form the very foundation of science. In comparison, "wiggles" have been left almost totally untouched.
>
> The goal of the present Essay is to face the challenge of building a Natural geometry of certain "wiggles," to be called "fractals."
>
> A most attractive feature of this plate is that it begs the scientist to "take the measure of the universe."[16]

The *regular* horizon says:

> Let us now turn back for another look at the science of geometry as a whole, to see what its starting-point is and how far it ranges from it, so as to get a view of the ordered cosmos of its ideas. Let us note that it is coextensive with all existing things, applies its reasoning to them all, and includes all their kinds in itself.... Always philosophizing about being in the manner of geometry, it has not only ideas but pictures of all the virtues—intellectual, moral, and physical—and presents in due order all the forms of political constitution, showing from its own nature the variety of the revolutions they undergo. In these areas its activity is immaterial and theoretical; but when it touches on the material world it delivers out of itself a variety of sciences...by which it benefits the life of mortals. Through these sciences it has...invented models for exhibiting the order of the whole heaven, and many things incredible to men it has unveiled and made credible to all.[17]

Moving between these two horizons, as they write themselves in our imaginations, we get our last view of the play or dance between purity and impurity, between the regular and the irregular. Do we see them embrace in their dance of the truth to space? Or do we see them break apart and dance in separate places here on this last page or screen, kept apart by the very thing that joins them: the space between them? Do we see the footprints, or the glazier?

NOTES

1. Benoit B. Mandelbrot, *The Fractal Geometry of Nature*, updated and augmented (New York: Freeman, 1983), p. 1. The book was first published in 1977.
2. Mandelbrot, *The Fractal,* p. 1.
3. Mandelbrot, *The Fractal*, p. 5.
4. Mandelbrot, *The Fractal*, p. 23.
5. Mandelbrot, *The Fractal*, p. 2.
6. Mandelbrot, *The Fractal*, p. 3.
7. Mandelbrot, *The Fractal*, p. 4.
8. Mandelbrot, *The Fractal*, p. 19.
9. Mandelbrot, *The Fractal*, p. 21.
10. Mandelbrot, *The Fractal*, p. 23.
11. Mandelbrot, *The Fractal*, p. 4.
12. Mandelbrot, *The Fractal*, p. 22.
13. Mandelbrot, *The Fractal*, pp. 22-23.
14. Mandelbrot, *The Fractal*, p. 23.
15. Mandelbrot, *The Fractal*, p. C2, between pp. 276-277. Mandelbrot's capitals.

16. Mandelbrot, *The Fractal*, p. C2, between pp. 276-277.

17. Proclus, *A Commentary on the First Book of Euclid's Elements*, trans. G. Morrow (Princeton, NJ: Princeton University Press, 1970), pp. 50-51. My italics.

Illustration 12. "Melancholia I," engraving, 1514, by Albrecht Dürer. (By permission of Victoria & Albert Museum, London/Art Resource, New York.)

Chapter 8

CONCLUDING CODA

The foregoing images and ideas of the geometric imaginary highlight some of the philosophical processes informing the action of "thinking geometrically." Naturally, many things were left unsaid, the foremost being the undeniable skill that geometers have in repeatedly holding in their imaginations complex shapes beyond the simple austere forms hanging in the gallery of traditional geometry. This ability, fielded through the question of imaging and illustrating figures "beyond the third dimension," yet being constantly bound by the current style of the world in any imagining of such figures, is a subject that deserves a wider and more informed study. I have also avoided dealing with the nature of specific geometric shapes and formulae, proofs, propositions, theorems, and the like, including the applied procedures whereby some proof is enacted. In the latter instances, to have done otherwise would have added nothing new to critical discourses that have in the past proceeded almost solely by offering new geometries. Since these have all issued forth from within the discourse of geometry itself, previous approaches have left in place the theoretical structure underpinning geometry as the arbiter of what will count as the truth of space as well as geometry as the sole arbiter of its own practice.

Another major form of critique that has grown in popularity of late, and that remains firmly within the purview of geometry itself, has centered upon what constitutes the foundational character of geometry and mathematics. It is captured in such questions as: "What is mathematics?" or "Why is mathematics so successful?" To ask in this fashion that the foundation of mathematics resolve itself toward the ethereal or the material is to harbor a redundant desire for certainty at an epistemological and ontological level. Therefore, in contrast to these questions and in contrast to discourse discreteness, this study has sought to cut its cloth from a different yard. My question has not been: What is mathematics?—as if mathematics had some original ontic validity which has been merely clouded over by applied success, or even lost in a narcissistic pool of self-gazing. For whatever pool geometry looks into, what it sees is always more than just its own reflection. Nor do we get very far if we continue to hold on to the Galilean idea that mathematics is the language in which nature is written. For what language is nature not written in?

Rather, my question has been: What is the character of geometry's inability ever to entirely separate itself from either what it is attempting to explain—namely, its own universality (say, idealities)—or what it is attempting to displace (say, nature, mercurialities)? Which is to say again, at the very dawn of the Greek miracle, geometry's horizon of immeasurability arose from within its horizon of measurability (just as the Greek miracle initially arose from within the surveyor's art of measuring the irregular). It arose from within itself as the outside other to itself. This arising, however, was wrought in advance as that which the horizon of measurability might possibly, one day, have need to redeem, or bring back into the folds of itself. The procedures for this "bringing back" were always there, but concealed through disavowal and displacement. In line, then, with this "hiding," it is now possible to advance a critique of geometry from outside its own point of view, so to speak, because of contemporary geometry's explicit experiential bias at the level of the formulation of its theories, coaligned with a certain overall crisis of identity characterizing much of science in our century.

Hence, in the shadow of geometry's current explicit interest in the experiential, I have sought to highlight the fact that, while the geometric imaginary is always in the style of the world, so, too, is the world always in the style of the geometric imaginary. This one thought—coextensive to its ally, the before-mentioned "arising"—underpins the idea of contamination as I have used this term. To be sure, this term has for many a decidedly negative sense. It need not, however, be taken exclusively in this way (as I hope has been indicated). In worrying this exclusivity, at the same time as leaving it in place, I have not sought to offer some neatly closed-off definition of the term (its edges will form without my help). I have, instead, let it gather its own names and airs as I have noted the processes staining what we take at any one moment to be the horizon of measure or unmeasure: that is, the double-sided screen of the geometric imaginary. In saying this, I recall that to tear one side of a piece of paper is always already to tear the other side.

In sum: The Greek miracle is not only the formulation of axiomatica. It is not just the emergence of a common world captured in a set of limited shapes and first principles. It is also the fact that geometry lays out in advance the very horizon of its miracles. But this is a horizon upon which the traces of a certain impurity may be also seen. This impurity, or footprints of an absent dancer, has of late been called back into service as the "new" purview of geometry. It is in this sense of a "calling back" that I have understood geometry as reversing its two-thousand-year-old binary. Although I have not elaborated on this, I sense that geometry will be successful at this reversal, if for no other reason than that geometry has already played such a formative role in how we have come to think of the world of the experiential as impure or irregular.

I will not, however, pretend that such a reversal, a "bringing" or "calling back" into the fold of what was *always* enfolded, has no alarming aspects. Foremost among these is dividing experience along mathematical-computer lines. However, against these bells of alarm, such a reversal also produces the opportunity for cultural theorists, theorists of the experiential, to lend their insights in the formation of this contemporary science of the mercurial. Mindful, they are themselves always embedded in an age-old theorization of the mercurial by geometry. In lieu, then, of more work in this promising field, the task standing before us, as science shifts its gears for the next millennium, is, as I see it, to continue to manipulate the contaminative flows between what it says at any one moment is the transcendent and what it says is the mercurial.

This shift in the interests of science leads to one more observation which, while derivable from the foregoing, was not pursued in any direct manner. Philosophies of science which ground their perceptions of past scientific change or development in radical paradigmatic breaks, or radical discontinuities, face severe difficulties. They fail to be able to explain how geometry, regardless of its many manifestations and systems, is in many senses always essentially geometry. It is, always, geometry *quâ* geometry. It continues unchallenged as the discourse that, through one form of measure or another, speaks the truth of space. Hence, the popular view that there is a clear and absolute distinction between traditional and modern geometry only makes sense if we disregard the fact that both periods employed the conception of axiomatica, or the idea of static lawful principles—even when some of these principles were later challenged and replaced, or viewed as special or local laws that were allowed to coexist with other first principles relating to a wider universe. I have in mind here Euclid's fifth postulate concerning parallel lines. (Even contemporary geometry gives every indication of retaining axiomatica, not as first principles but rather as essential processes: fixed idealities replaced by mobile mercurialities.)

However, while there is no point in denying that there are profound differences in the interests of various scientific practices throughout their long histories, any more than in trying to deny that any particular period also manifests a broad set of shared interests or positions, it is something else again to argue that these differences can be thought of entirely in some absolutely differentiated identity. We have to let go of the possibility of ever being able to think of the unity of science in the fullness of some undifferentiated totality. The task, as I understand it, is to come to a comprehension of the processes permitting us to think of differences as-if they were uniquely separate, at the same time as we seek to think of unity as-if it was some undifferentiated totality. It is here, in this tension, that geometry as a generator of meaning is a convenient way to think the as-if-ness of separation and the as-if-ness of unity—not least because geometry, if it is any one thing at all, is certainly the process of differentiating through measure, at the same moment as it seeks, through measure, to unify these self-same differences.

Illustration 13. Illustration from *Libro d'Antonio Labacca appartenente a l'architettura*, nel qual si figurano alcune notabili antiquita di Roma (Rome: Lafreri, 1568). Antonio Labacca was born circa 1495. (Reproduced courtesy of the Rare Books Room, Pennsylvania State University Libraries.)

NOTE ON THE ILLUSTRATION OF PLATO'S ALLEGORY OF THE CAVE

P. Hans Sun and James A. Lynch

Remarkably, despite its fame and influence, Plato's allegory of the cave has seldom been illustrated. The 1604 engraving "Antrum Platonicum" by the Dutch artist Jan Saenredam (c. 1565-1607) is a rare exception.

The justly famous allegory occurs in Plato's *Republic* at the opening of book VII, immediately after what is commonly referred to as the myth of the divided line which closes book VI. The idea that this life may be described as like living in an underworld perhaps has anticipations in Empedocles,[1] and its origins have been variously examined.[2] Its influence may be found in philosophical texts as diverse as Aristotle and Cicero,[3] Francis Bacon,[4] George Berkeley,[5] and Carl Jung,[6] as well as in numerous literary references.[7]

The allegory is presented to Glaucon by Socrates in four movements. In the first movement, Socrates sets the stage, as it were, by asking Glaucon to:

> make an image of our nature [*phusis*] in its education [*paideia*] and want of education, likening it to a condition of the following kind. See human beings as though they were in an underground [*katagaios*] cavern-like [*spelaiodes*] dwelling [*oikesis*]...with their legs and necks in bonds so that they are fixed, seeing only in front of them, unable...to turn their heads all the way around. Their light is from a fire burning far above and behind them. Between the fire and the prisoners there is a road above, along which see a wall, build like the partitions puppet-handlers set in front of the human beings and over which they show the puppets....Then also

> see along this wall human beings carrying all sorts of apparatus [*skeue*] which project above the wall, and statues of human beings and other animals (514a-c).[8]

Such human beings, Glaucon agrees, would see only "the shadows cast by the fire on the opposite side [*katantikru*] of the cavern" (515a).

In the second movement, Socrates asks Glaucon to consider what might result if "by nature" the prisoners were released from their bonds and compelled to get up, turn around, and look up at the light.

In the third movement, Socrates asks Glaucon to consider further what would happen if someone were dragged out of the cavern home and into the sunlight:

> Then I suppose he would have to become accustomed, if he were going to see what is up above. At first he would most easily make out the shadows; and after that the phantoms of human beings and other things in [a pool of] water; and, later, the things themselves. And from there he could turn to beholding the things in heaven and heaven itself, more easily at night, looking at the light of the stars and the moon, than by day, looking at the sun and sunlight (516a-b).

In this allegory, there are two things that function as what Noël Gray has referred to as screens: the cavern wall and a pool, stone and water. It is no doubt significant that on stone only shadows fall, shadows constructed by an absence of light, whereas in the water one detects richer images composed on (weakened) light. Interestingly enough, the history of photography traces its origins back to the shadows on the wall, but not to the reflections in the pool of water.[9]

Saenredam's etching presents only the first two movements in the allegory and provides no indication of the pool of water standing outside the cave.

With regard to Saenredam's etching itself: above the image is written "Lux venit in mundum et dilexerunt homines magis tenebras quam lucem" (Light has come into the world and humans have loved darkness more than light—John 3:19). In the Revised Standard translation of the Bible, the passage is rendered as, "And this is the judgment, that the light has come into the world, and people loved darkness rather than light because their deeds were evil."

The extended passage below the image reads:

> Maxima pars hominum caecis immersa tenebris
> volvitur assidue, et studio laetatur inani.
> Adspice ut obiectis obtutus inhaereat umbris:
> ut veri simulacra omnes mirentur amentque,
> et stolidi vana ludantur imagine rerum.
> Quam pauci meliore luto, qui in lumine puro,
> secreti a stolida turba, ludibria cernunt,
> rerum umbras, rectaque expendunt omnia lance.
> Hi, posita erroris nebula, dignoscere possunt
> vera bona, atque alios caeca sub nocte latentes
> extrahere in claram lucem conantur; at illis
> nullus amor lucis: tanta est rationis egestas.

A literal translation may be rendered as follows:

> The greatest part of humanity, plunged into blind darkness,
> ponders incessantly and rejoices in vain studies.
> Behold, the gaze fixes upon the presented abundance:
> so that all love and marvel at simulacra of truth,
> and the foolish play with the empty images of things.
> Few are better than clay, who in the pure light,
> separated from the dull crowd, understand the joke,
> the shadows of things, and weigh all with the proper scale.
> They are able to distinguish within the fog of error
> the true good, and even attempt to draw out into the clear light
> others concealed in the blindness of night: but they
> have no love for the light: such is the poverty of reason.

Beneath this passage is the following: "H.L. Spiegel figurari et sculpi curavit ac doctissimo ornatissimoque domino Petro Paas in Lugdunensi Academica professori medico donum dedit" (H.L. Spiegel commissioned these figures and engravings and gave them as a present to the most elaborately learned lord Petro Paas, professor of medicine in the Lugdenensi Academica [Academy of Lyon]).

The engraving thus appears related to the work of Henric Laurenszoon Spieghel (1549-1612), a poet of the northern Dutch Renaissance. Spieghel was a well educated Catholic in an area that had become Protestant. His greatest poem, *Hertspiegel* (Heart-mirror; 1614), is a long allegory that employs a religious conflation of Platonism and Christianity to emphasize the centrality of the virtuous life over dogma. It was Spieghel's use of Plato's allegory in *Hertspiegel* (book III, lines 29-99), that inspired the artist Cornelisz van Haarlem (1562-1638) to do a drawing of Plato's cave which is no longer extant but which served in turn as the basis for Saenredam's engraving.

This unique rendering of Plato's allegory of the cave under the aegis of a Johannine Christian theology of light has been examined at length by Philippe Verdier, who points out many of its special features.[10] An insufficiently recognized contrast, however, would note a fundamental ambiguity in the cultural interpretation of the cave experience. As Socrates tells the story to Glaucon, to ascend from the cave is to approach closer to the source of reality, from the rock screen of shadows to the liquid screen of reflections. Yet, in a view to be found in as diverse sources as those of Mediterranean mystery religions and Feodor Dostoevsky's *Notes from the Underground* (1864), it is within the depths of the earth, sheltered from the dominating lights of the everyday, that truth and reality are to be encountered.

The screen becomes, as it were, a deeper reality, and shadows a testimony to solidity—a reversal on which art historian E.H. Gombrich has commented as follows:

> In the West we have the famous story by Adelbert von Chamisso, first published in 1814, of the unhappy Peter Schlemihl who is persuaded by the devil to sell his shadow, with dire consequences. Since he casts no shadow he has lost his place in the real world.

While in that story the presence of the shadow indicates solid reality, one of the most famous philosophical parables in the Western tradition draws the opposite conclusion. I refer to the passage in Plato's Republic where he compares the human condition to that of prisoners so fettered in a cave that they can only see the wall opposite its opening. Consequently they come to regard the shadows cast by the events outside the wall of the cave as reality till they are enabled to turn round and to recognize their error. The engraving by Saenredam gives an elaborate Christianized interpretation of the passage, singling out some of the features Plato described such as the bearers on the wall carrying "all sorts of articles...made of stone or wood," which are here turned into images of personifications....

Plato reminded his readers of the tricks of shadow players or puppet shows, for in fact the philosopher looked askance at the illusions created by painters, notably the scene-painters who managed to deceive the senses and to present us with a phantom world. It appears that the Greek term used for illusionistic painting was in fact skiagraphia, shadow painting, though it is hard to decide whether the term implied the actual rendering of shadows or merely the use of light and shade for the purpose of modelling....It was this latter device that revolutionized the art of painting and was destined to become the distinguishing mark of the Western tradition. Neither the art of Egypt nor Greek vase paintings of the early fifth century show signs of modelling or shading, but after the invention of this technique it was never wholly discarded in the West. It was different with cast shadows, which seem to have come and gone, very much like our shadows when walking along a lamplit road.[11]

In classical literature, no one was perhaps more aware of this fundamental ambiguity of the cave allegory than Cervantes. Book II of his *Quixote* has several related episodes in which, however, true to his habits, he modifies this source material and provides us with an allegory as viewed from within a house of mirrors. In chapters 22 and 23, Don Quixote sets off on an expedition to the enchanted Cave of Montesinos which, tellingly, has been connected in Spanish folklore with the nearby twin lakes of Ruidera. Quixote has himself lowered into the cave so that he might see the marvels that it contains, but an hour later his companions pull him out only to find him fast asleep. When he awakens, he relates how he had spent three days underground in the company of various enchanted knights and ladies. His companion, Sancho, expresses his doubts about the account, stating:

It was an evil moment, a worse season, and a sorrowful day when your mercy, my dear patron, descended into tho other world, and an evil happening when you met with Sr. Montesinos, who has returned you to us in this way. You were fine above ground with all of your faculties, which God had given you, speaking words of wisdom and giving advice at each step, but not now, since you speak the biggest absurdities that can be imagined (II, ch. 23).

In Cervantes' version, therefore, Quixote takes on the role of the liberated cave dweller who returns to reveal the truth to his unbelieving companions. But, interestingly, the locations of the sensible and intelligible worlds are reversed: the "realm of the senses" is, for Cervantes, also the realm of daylight, while the "realm of reason" is actually the underground cavern. Or are they? Surely, the don did not indeed visit an enchanted realm in the cave. At the beginning of the following chapter, the narrator himself sees fit to express his doubts, while not arriving at any definite conclusions:

> If this adventure seems apocryphal, I am not to blame; and so, without professing it to be false or true, I write it. You, reader, since you are prudent, judge it as it appears to you, for I dare not and cannot do more (II, ch. 24).

Quixote himself defends his tale, however, by asserting, "The passing of time will tell, Sancho...since time, which uncovers all things, leaves nothing that it does not pull out into the light of the sun, even though it was hidden in the bosom of the earth" (II, 25).

Immediately after this episode, we find another which shows that, if Quixote had found it difficult to live in the sensible world before his descent into the realm of reason, he is even less able to do so now. While watching a puppet show (reminiscent of Plato's shadow show) about a Christian princess fleeing from a Moorish army, he rises to defend her and slashes the Moorish puppets to pieces with his sword. Coming to his senses, he exclaims:

> Now I have just realized...what I have believed many times before: these enchanters who pursue me simply put figures like these before my eyes, and then change them into whatever they want them to be. Really and truly I tell you gentlemen listening to me, that to me everything that has happened here has seemed to be real (II, ch. 26).

Once again being fooled by the shadows, Quixote has become a "useless" philosopher whose blindness, caused by his return to the sensible world, is the most pitiable kind.

Finally, as if following Don Quixote, the world constituted by the screens of the electric and electronic media, from film to television and computer, has been described as a return to Plato's cave—but with a difference. Media critic John O'Neill, for instance, presents:

> the imagery of Plato's cave [as] a challenge to contemporary self-understanding. For we cannot escape into a higher metaphysics or into an elite culture that will exclude more than half of humanity in a self-inflicted wound of pride and privilege.[12]

For O'Neill and related interpreters, electronic screens no longer simply reflect—as do stone and water—but shine forth with their own phenomena which, although in some sense unreal, are not to be abandoned.

NOTES

1. In a series of fragments that appear to describe a cycle of incarnations, Empedocles writes, "We came under this roofed cave [*antron*]" (frag. 120). The passage in preserved in Porphyry's *De antro nympharum* (8). A disciple of Plotinus, Porphyry's work is itself a short commentary on a passage near the beginning of Book 13 of the *Odyssey*, where Homer narrates the story of Odysseus' entrance into the cave of the Naiades. In Porphyry as well as in Plotinus, *Enneads* (IV, viii, 1), where both Empedocles' cave and Plato's are mentioned, there is an assumption that the two are related. In Porphyry's words, it is simply quite natural that the ancients took the cave as a symbol of the world and

its invisible powers because, "as caves are obscure, so is the essence of these powers" (*De antro nympharum*, 7).

2. See, especially, John Henry Wright, "The Origin of Plato's Cave," *Harvard Studies in Classical Philology*, vol. 17 (1906), pp. 131-142. Wright's theory is that Plato's allegory is most likely to be associated with a geographical cave near the village of Vari on Mount Hymettus.

3. See Cicero, *De natura deorum* (II, 37), who quotes Aristotle as having written in his lost dialogue "On Philosophy": "If there were beings who had always lived beneath the earth, in fine, well-lit dwellings, decorated with works of art and pictures and furnished with all the luxuries enjoyed by persons thought to be supremely happy, and who though they had never come forth above the ground had learnt by report and by hearsay of the existence of certain deities or divine powers; and then if at some time the jaws of the earth were opened and they were able to escape from their hidden abode and to come forth into the regions which we inhabit; when they suddenly had sight of the earth and the seas and the sky, and came to know of the vast clouds and mighty winds, and beheld the sun, and realized not only its magnitude and beauty but also its efficacy in causing the day by diffusing light over all the sky, and, after night had darkened the earth, they then saw the whole sky spangled and adorned with stars, and the changing phases of the moon's light, waxing and waning, and the risings and settings of all these heavenly bodies and their courses fixed and changeless throughout all eternity—when they saw these things, surely they would think that the gods exist and that such would be their work." (This English translation is a slight modification of that by H. Rackham in the Loeb Classical Library series, 1933.)

4. Francis Bacon, *Novum Organum* (1620, book I), has an extended discussion of the "Idols of the Cave." See especially aphorisms 53 ff.

5. George Berkeley, *Siris* (1744, para. 263): "It cannot be denied that, with respect to the universe of things, we in this mortal state are like men educated in Plato's cave, looking on shadows with our backs turned to the light."

6. Carl Jung, *Analytical Psychology: Notes of the Seminar Given in 1925*, ed. William McGuire (Princeton, NJ: Princeton University Press, 1989), p. 232.

7. See, for example: Henry More, *Psychathanasia* (1647, book I, canto i, 10): "Like men new made contriv'd into a cave/That ne're saw light, but in that shadow pit." Sir Thomas Browne, *Hydriotaphia or Urn Burial* (1658, ch. 4): "A dialogue between two infants in the womb concerning the state of this world, might handsomely illustrate our ignorance of the next, whereof methinks we yet discourse in Plato's den, and are but embryo philosophers" (*Sir Thomas Browne's Works*, ed. Simon Wilkin [1835], vol. 3, p. 486).

8. The quotations are adapted from *The Republic of Plato*, trans. Allan Bloom (New York: Basic Books, 1968).

9. See, for example, Robert Rosenblum, "The Origin of Painting: A Problem in the Iconography of Romantic Classicism," *Art Bulletin*, vol. 39 (December 1957), pp. 279-290. Another story is that of a Corinthian maid (often named Dibutade) who traced the outline of her lover's shadow to create a permanent reminder of him before he left for an extended period in another country.

10. Philippe Verdier, Des mystères grecs à l'âge baroque. Commentaires à l'«Antrum Platonicum» de J. Saenredam," in Antje Kosegarten and Peter Tigler, eds., *Festschrift Ulrich Middeldorf* (Berlin: Walter De Gruyter, 1968), vol. 1, pp. 376-391, and vol. 2, plate 174. See also the much more brief commentary in Konrad Gaiser, *Il paragone della caverna: Variazioni da Platone a oggi* (Naples: Bibliopolis, 1985), especially pp. 42-43 and 82-84.

11. E.H. Gombrich, *Shadows: The Depiction of Cast Shadows in Western Art* (London: National Gallery Publications, 1995), pp. 17-19. This volume was written to accompany an exhibition at the National Gallery in London, April 26-June 18, 1995. See also Machael Baxandall, *Shadows and Enlightenment* (New Haven, CT: Yale University Press, 1995).

12. John O'Neill, *Plato's Cave: Desire, Power, and the Specular Functions of the Media* (Norwood, NJ: Ablex, 1991), p. 7. See also Jay Newman, *The Journalist in Plato's Cave* (Rutherford, NJ:

Fairleight Dickinson University Press, 1989); and Alvin B. Kernan, *In Plato's Cave* (New Haven, CT: Yale University Press, 1999).

Illustration 14. Orpheus being killed by the Baccantes, with his dying hand breaking the triangle of one of twelve vaulting cells in Andrea Mantegna's *Camera degli Sposi* (Marriage chamber) of 1474, often called "the most beautiful room in the world." Drawing by Carlos Verdadero.

BIBLIOGRAPHY

The following references supplement those present in the text and notes. They are divided into three main categories—science (including the the philosophy of science), philosophy, and aesthetics—reflecting the primary themes of this study. The bibliography also adds a brief list of general references.

SCIENCE

Albarn, Keith, and J. Smith. 1977. *Diagram: The Instrument of Thought.* London: Thames and Hudson.

Aleksander, Igor. 1982. *Artificial Vision for Robots.* New York: Kogan Page.

Alexander, H. 1956. *The Leibniz-Clarke Correspondence, Together with Extracts from Newton's Principia and Optiks.* Manchester, UK: Manchester University Press.

Almgren, Frederick, and J. Taylor. 1976. "The Geometry of Soap Films and Soap Bubbles," *Scientific American*, vol. 235, no. 1 (January), pp. 82-93.

Almgren, Frederick, and J. Sullivan. 1992. "Visualizations of Soap Bubble Geometries," *Leonardo*, vol. 25, nos. 3 and 4, pp. 267-271.

Armitage, Angus. 1947. *The World of Copernicus.* New York: Signet.

Aspray, William, and Philip Kitcher, eds. 1988. *Minnesota Studies in the Philosophy of Science,* Vol. 11: *History and Philosophy of Mathematics.* Minneapolis, MN: University of Minnesota Press.

Bachelard, Gaston. 1984. *The New Scientific Spirit.* Arthur Goldhammer, trans. Boston: Beacon Press.

Barnsley, Michael. 1988. *Fractals Everywhere.* Boston: Academic Press.

Beck, William. 1957. *Modern Science and the Nature of Life.* New York: Penguin.

Benedikt, Michael, ed. 1991. *Cyberspace: First Steps.* Cambridge, MA: MIT Press.

Bohm, David. 1951. *Quantum Theory.* Englewood Cliffs, NJ: Prentice-Hall.

Bohm, David. 1957. *Causality and Chance in Modern Physics.* London: Routledge and Kegan Paul.

Bohm, David. 1965. *The Special Theory of Relativity.* London: Benjamin.

Bohm, David. 1980. *Wholeness and the Implicate Order.* London: Routledge and Kegan Paul.

Bohm, David, and F. David Peat. 1987. *Science, Order, and Creativity.* Toronto: Bantam.

Boyd, R., P. Gasper, and J. Trout, eds. 1991. *The Philosophy of Science.* Cambridge, MA: MIT Press.

Boyer, Carl Benjamin. 1991. *A History of Mathematics.* New York: John Wiley.

Briggs, John, and F. David Peat. 1989. *Turbulent Mirror: An Illustrated Guide to Chaos Theory and the Science of Wholeness.* New York: Harper and Row.

Burchett, Richard. 1856. *Linear Perspective*. London: Chapman and Hall.

Callahan, J.J. 1976. "The Curvature of Space in a Finite Universe," *Scientific American*, vol. 235, no. 2 (February), pp. 90-100.

Capra, Fritjof. 1982. *The Turning Point: Science, Society and the Rising Culture*. London: Fontana.

Chambers, D. 1984. *Putting Nature in Order*. Victoria, Australia: Deakin University.

Danto, Arthur, and Sidney Morgenbesser, eds. 1960. *Philosophy of Science: Science, Language, and Experience; Law and Theories, Space, Time, and Causality*. New York: New American Library.

Daston, Lorraine, and Peter Galison. 1992. "The Image of Objectivity," *Representations*, vol. 40, pp. 81-128.

Davis, Philip, and Reuben Hersch. 1981. *The Mathematical Experience*. Harmondsworth, Middlesex, UK: Penguin.

Dear, Peter. 1995. *Discipline and Experience: The Mathematical Way in the Scientific Revolution*. Chicago: University of Chicago Press.

Deken, Joseph. 1983. *Computer Images: State of the Art*. London: Thames and Hudson.

Edgerton, Samuel. 1991. *The Heritage of Giotto's Geometry Art and Science on the Eve of the Scientific Revolution*. Ithaca, NY: Cornell University Press.

Ekeland, Ivar. 1988. *Mathematics and the Unexpected*. Chicago: University of Chicago Press.

Elkins, James. 1992. "On Visual Desperation and the Bodies of Protozoa," *Representations*, vol. 40, pp. 33-56.

Fenwick, Peter, and David Lorimer.1989. "Can Brains be Conscious?" *New Scientist*, vol. 123 (August 5), pp. 54-56.

Fleischmann, M., et al, eds. 1990. *Fractals in Natural Sciences from the Proceedings of the Royal Society of London*. Princeton, NJ: Princeton University Press.

Friedhoff, Richard, and William Benzon. 1988. *The Second Computer Revolution: Visualization*. New York: W.H. Freeman.

Galileo. 1967. *Dialogue Concerning the Two Chief World Systems—Ptolemaic and Copernican*. Stilman Drake, trans. Berkeley, CA: University of California Press.

Gans, David. 1973. *An Introduction to Non-Euclidean Geometry*. New York: Academic Press.

Ghyka, Matila. 1977. *The Geometry of Art and Life*. New York: Dover.

Gleick, James. 1987. *Chaos: Making a New Science*. New York: Viking.

Gray, Noël. 1995. "Geometry of Constructions, Constructions of Geometry," in *Research in Philosophy and Technology*, vol. 15, ed. C. Mitcham. Greenwich, CT: JAI Press, 1975, pp. 21-36.

Gray, Noël. 1994. "Geometry and the Sublime: Imagination and the Closure of Creativity," in "*The Most Sublime Act*". *Essays on the Sublime*, eds. T. Rachwal and T. Slawek. Katowice, Poland: Wydawnictwo Uniwersytetu Slaskiego, pp. 9-19.

Gray, Noël. 1994. "Seeing Nature," in *The Construction of Nature: A Discursive Strategy in Modern European Thought*, eds. S. Grgas and S. Larsen. Denmark: Odense University Press, pp. 211-223.

Gray, Noël. 1994. "The Mathematization of Experience in Virtual Realities," in *Selected Papers of 1992 European Integration and the European Mind Third Conference of the International Society for the Study of European Ideas*, eds. E. Talmor and S. Talmor. Oxford, UK: Pergamon Press, 1994); also in *History of European Ideas*, vol. 20, nos. 1-3, pp. 341-348.

Gray, Noël. 1991. "Reading The Face," in *Bodies in Conflict*. Sydney, Australia: Tin Sheds Art Gallery and Workshop, Department of Fine Arts and Department of Architecture, University of Sydney, pp. 34-39.

Gray, Noël. 1991. "Critique and a Science for the Sake of Art: Fractals and the Visual Arts (A Response to the Aesthetic Theory of Benoit Mandelbrot and Fractal Geometry [*Leonardo*, 1989])," *Leonardo*, vol. 24, no. 3 (1991), pp. 317-321.

Greenberg, Marvin. 1972. *Euclidean and Non-Euclidean Geometry, Development and History*. San Francisco: W.H. Freeman.

Hall, Edward. 1969. *The Hidden Dimension.* New York: Anchor.

Hall, Nina, ed. 1992. *The New Scientist Guide to Chaos.* New York: Penguin. See especially: I. Percival, "Chaos: A Science for the Real World," pp. 11-21; I. Stewart, "Portrait of Chaos," pp. 44-58; B. Mandelbrot, "Fractals—A Geometry of Nature," pp. 122-135; C. Series, "Fractals, Reflections and Distortions," pp. 136-148; J. Lesurf, "Chaos on the Circuit Board," pp. 162-173; M. Berry, "Quantum Physics on the Edge of Chaos," pp. 184-195; P. Coveney, "Chaos, Entropy and the Arrow of Time," pp. 201-212; P. Davies, "Is the Universe a Machine?" pp. 213-221.

Haugeland, John. 1985. *Artificial Intelligence: The Very Idea.* Cambridge, MA: MIT Press.

Hayles, N. Katherine, ed. 1991. *Chaos and Order: Complex Dynamics in Literature and Science.* Chicago: University of Chicago Press.

Heath, Thomas, trans. 1908. *The Thirteen Books of Euclid's Elements.* 2 vols. Cambridge, UK: Cambridge University Press.

Hofstadter, Douglas. 1979. *Gödel, Escher, Bach.* New York: Vintage Books.

Horgan, John. 1992. "'The Lonely Odysseus of Particle Physics'—Profile: Murray Gell-Mann," *Scientific American*, vol. 226, no. 3 (March), pp. 30-32.

Kerszberg, Pierre. 1992. "Of Exact and Inexact Essences in Modern Physical Science," in *Phenomenology of Natural Science*, eds. Lee Hardy and Lester Embree. Dortrecht, The Netherlands: Kluwer, pp. 93-118.

Kline, Morris. 1959. *Mathematics and the Physical World.* London: John Murray.

Kline, Morris. 1972. *Mathematical Thought from Ancient to Modern Times.* New York: Oxford University Press.

Kuhn, Thomas. 1970. *The Structure of Scientific Revolutions*, 2nd ed. Chicago: Chicago University Press.

Kurzweil, Raymond. 1990. *The Age of Intelligent Machines.* Cambridge, MA: MIT Press.

Lakatos, Imre, and Alan Musgrave, eds. 1970. *Criticism and the Growth of Knowledge.* Cambridge, UK: Cambridge University Press.

Lynch, Michael, and Steve Woolgar, eds. 1990. *Representation in Scientific Practice.* Cambridge, MA: MIT Press.

Mandelbrot, Benoit. 1990. "Fractal Geometry: What Is It, and What Does It Do?" in Fleischman et al., eds., pp. 3-16.

Mandelbrot, Benoit. 1989. "Fractals and an Art for the Sake of Science," *Leonardo* Supplemental Issue: *Computer Art in Context*, vol. 22, pp. 21-24.

Markus, G. 1992. "Changing Images of Science," *Images of Science, Conditions of Democracy (Thesis Eleven)*, no. 33, pp. 1-56.

May, Robert. 1989. "The Chaotic Rhythms of Life," *New Scientist*, vol. 124 (November 18), pp. 37-41.

Metelli, Fabio. 1974. "The Perception of Transparency," *Scientific American*, vol. 230, no. 4 (April), pp. 90-98.

Miller, Arthur. 1987. *Imagery in Scientific Thought Creating 20th-Century Physics.* Cambridge, MA: MIT Press.

Mitcham, Carl. 1994. *Thinking through Technology: The Path Between Engineering and Philosophy.* Chicago: University of Chicago Press.

Monad, Jacques. 1971. *Chance and Necessity: An Essay on the Natural Philosophy of Modern Biology.* New York: Vintage Books.

Mullin, Tom. 1989. "Turbulent Times for Fluids," *New Scientist*, vol. 124 (November 11), pp. 52-55.

Murray, Carl. 1989. "Is the Solar System Stable?" *New Scientist*, vol. 124 (November 25), pp. 60-63.

Newman, John von. 1956. *The World of Mathematics.* 4 vols. New York: Simon and Schuster.

Nicolson, Marjorie. 1976. *Science and Imagination.* Hamden, CT: Archon.

Palmer, Tim. 1989. "A Weather Eye on Unpredictability," *New Scientist*, vol. 124 (November 11), pp. 56-59.

Peitgen, Heinz-Otto, and Peter Richter, eds. 1986. *The Beauty of Fractals: Images of Complex Dynamical Systems*. Berlin: Springer-Verlag. See especially: H. Peitgen and P. Richter, "Frontiers of Chaos," pp. 1-20; B. Mandelbrot, "Fractals and the Rebirth of Iteration Theory," pp. 151-160; A. Douady, "Julia Sets and the Mandelbrot Set," pp. 161-173; G. Eilenberger, "Freedom, Science, and Aesthetics," pp. 175-180; H. Franke, "Refractions of Science into Art," pp. 181-187.

Percival, Ian. 1989. "Chaos Reigns: 'Chaos: A Science for the Real World'," *New Scientist*, vol. 124 (October 21), pp. 42-47.

Peterson, Ivars. 1988. *The Mathematical Tourist: Snapshots of Modern Mathematics*. New York: W.H. Freeman.

Prigogine, Ilya, and Isabelle Stengers. 1984. *Order Out Of Chaos: Man's New Dialogue with Nature*. London: Flamingo.

Schroeder, Manfred. 1991. *Fractals, Chaos, Power Laws: Minutes from an Infinite Paradise*. New York: W.H. Freeman.

Scott, Stephen. 1989. "Clocks and Chaos in Chemistry," *New Scientist*, vol. 124 (December 2), pp. 53-59.

Shanker, Stuart, ed. 1988. *Gödel's Theorem in Focus*. London: Routledge.

Sklar, Lawrence. 1985. *Philosophy and Spacetime Physics*. Berkeley, CA: University of California Press.

Smart, John, ed. 1964. *Problems of Space and Time*. New York: Macmillan.

Stewart, Ian. 1989. "Portrait of Chaos," *New Scientist*, vol. 124 (November 4), pp. 42-47.

Tufte, Edward R. 1983. *The Visual Display of Quantitative Information*. Cheshire, CT: Graphics Press.

Van Helden, Albert. 1985. *Measuring The Universe: Cosmic Dimensions from Aristarchus to Halley*. Chicago: University of Chicago Press.

Vivaldi, Franco. 1989. "An Experiment with Mathematics," *New Scientist*, vol. 124 (October 28), pp. 46-49.

Wood, James. 1801. *Elements of Optics, Designed for the Use of Students in the University*. Cambridge: J. Burgess.

PHILOSOPHY

Aristotle. 1941. *The Basic Works of Aristotle*. Richard McKeon, ed. New York: Random House.

Aristotle. 1986. *De Anima [On The Soul]*. H. Lawson-Tancred, trans. New York: Penguin.

Atherton, Margaret. 1990. *Berkeley's Revolution in Vision*. London: Cornell University Press.

Barnes, Jonathan. 1975. "Aristotle's Theory of Demonstration," in *Articles on Aristotle*, vol. 1, eds. Malcolm Schofield, Richard Sorabji, and Jonathan Barnes. London: Gerald Duckworth, pp. 65-87.

Barnes, Jonathan. 1987. *Early Greek Philosophers*. New York: Penguin.

Bachelard, Gaston. 1964. *The Psychoanalysis of Fire*. C. Ross, trans. London: Quartet Encounters.

Bachelard, Gaston. 1987. *On Poetic Imagination and Reverie*. Colette Gaudin, trans. Dallas, TX: Spring.

Bathrick, David et al, eds. 1984. *German Critique: Modernity and Postmodernity*. No. 33 (Fall 1989).

Baynes, Kenneth, et al, eds. 1987. *After Philosophy, End or Transformation?* Cambridge, MA: MIT Press.

Bello, A., ed. 1981. *Analecta Husserliana: The Yearbook of Phenomenological Research, The Great Chain of Being, and Italian Phenomenology*. Holland: D. Reidel.

Bergson, Henri. 1960. *Creative Evolution*. A. Mitchell, trans. London: MacMillan.

Brumbaugh, Robert. 1964. *The Philosophers of Greece*. London: George Allen and Unwin.

Bubner, Rüdiger. 1981. *Modern German Philosophy*. London: Cambridge University Press.

Butcher, Samuel. 1951. *Aristotle's Theory of Poetry and Fine Art with a Critical Text and Translation of "The Poetics."* New York: Dover.

Cassirer, Ernst. 1953. *The Philosophy of Symbolic Forms*. 2 vols. R. Manheim, trans. New Haven, CT: Yale University Press.

Cornford, Francis. 1935. *Plato's Theory of Knowledge: The Theaetetus and the Sophist of Plato*. London: Kegan Paul.

Crombie, I.M. 1963. *An Examination of Plato's Doctrines*. 2 vols. London: Routledge and Kegan Paul.

De Man, Paul. 1972. "Genesis and Genealogy in Nietzsche's *The Birth of Tragedy*," *Diacritics*, vol. 11, no. 4 (Winter), pp. 44-53.

Derrida, Jacques. 1978. *Writing and Difference*. A. Bass, trans. Chicago: University of Chicago Press.

Derrida, Jacques. 1987. *The Truth In Painting*. G. Bennington and I. McLeod, trans. Chicago: Chicago University Press.

Feyerabend, Paul. 1975. *Against Method: Outline of an Anarchistic Theory of Knowledge*. London: Verso.

Gasché, Rodolphe. 1986. *The Tain of the Mirror: Derrida and the Philosophy of Reflection*. Cambridge, MA: Harvard University Press.

Glacken, Clarence. 1967. *Traces on the Rhodian Shore: Nature and Culture in Western Thought from Ancient Times to the End of the Eighteenth Century*. Berkeley, CA: University of California Press.

Gurwitsch, Aron. 1949-1950. "Gelb-Goldstein's Concept of 'Concrete' and 'Categorical' Attitude and the *Phenomenology of Ideation*," *Philosophy and Phenomenological Research*, vol. 10 (September 1949—June 1950), pp. 172-196.

Gurwitsch, Aron. 1985. *Marginal Consciousness*. Lester Embree, ed. Athens, OH: Ohio University Press.

Hartman, Geoffrey. 1981. *Saving the Text: Literature, Derrida, Philosophy*. Baltimore: Johns Hopkins University Press.

Husserl, Edmund. 1960. *Cartesian Meditations*. D. Cairns, trans. The Hague: Martinus Nijhoff.

Husserl, Edmund. 1964. *The Idea of Phenomenology*. W. Alston, M. Nijhoff, and G. Nakhnikian, trans. The Hague, The Netherlands: Martinus Nijhoff.

Husserl, Edmund. 1973. *Experience and Judgement: Investigations in a Genealogy of Logic*. J. Churchill and K. Ameriks, trans. Evanston, IL: Northwestern University Press.

Kant, Immanuel. 1883. *Prolegomena and Metaphysical Foundations of Natural Science*. E. Bax, trans. London: George Bell.

Kant, Immanuel. 1956. *Critique of Practical Reason*. L. Beck, trans. New York: Macmillan.

Kant, Immanuel. 1977. *Prolegomena: To Any Future Metaphysics that Will Be Able to Come forward as Science*. P. Carus, trans. Revised by J. Ellington. Indianapolis, IN: Hackett.

Körner, Stephan. 1969. *Fundamental Questions of Philosophy*. Sussex, UK: Harvester Press.

Lyotard, Jean. 1986. *Phenomenology*. Brian Beakely, trans. Albany, NY: State University of New York Press.

Mortley, Raoul. 1991. *French Philosophers in Conversation: Levinas, Schneider, Serres, Irigaray, LeDoeuff, Derrida*. London: Routledge.

Natanson, Maurile. 1973. *Edmund Husserl: Philosopher of Infinite Tasks*. Evanston, IL: Northwestern University Press.

Nietzsche, Friedrich. 1976. *Beyond Good and Evil: Prelude to a Philosophy of the Future*. R. Hollingdale, trans. London: Penguin Books.

Nietzsche, Friedrich. 1977. *A Nietzsche Reader*. R. Hollingdale, trans. London: Penguin Books.

Olson, Richard. 1975. *Scottish Philosophy and British Physics 1750-1880: A Study in the Foundations of the Victorian Scientific Style*. Princeton, NJ: Princeton University Press.

Plato. 1989. *The Collected Dialogues of Plato, Including the Letters*. E. Hamilton and H. Cairns, trans. Princeton, NJ: Princeton University Press.

Proclus. 1987. *Commentary on Plato's* Parmenides. G. Morrow and J. Dillon, trans. Princeton, NJ: Princeton University Press.

Raven, John. 1965. *Plato's Thought in the Making: A Study of the Development of His Metaphysics*. Cambridge, UK: Cambridge University Press.

Reichenbach, Hans. 1958. *The Philosophy of Space and Time*. M. Reichenbach and J. Freund, trans. New York: Dover.

Serres, Michel. 1989. "Panoptic Theory," in *The Limits of Theory*, ed. T. Kavanagh. Stanford, CA: Stanford University Press, pp. 25-47.

Snell, Bruno. 1982. *The Discovery of the Mind in Greek Philosophy and Literature*. New York: Dover.

Sokolowski, Robert. 1974. *Husserlian Meditations: How Words Present Things*. Evanston, IL: Northwestern Press.

Stott, Douglas W. 1989. *F.W.J. Schelling: The Philosophy of Art*. Minneapolis, MN: University of Minnesota Press.

Versényi, Laszlo. 1963. *Socratic Humanism*. New Haven, CT: Yale University Press.

Walker, Ralph. 1978. *Kant: The Arguments of the Philosophers*. London: Routledge and Kegan Paul.

Werkmeister, William Henry. 1980. *Kant: The Architectonic and Development of His Philosophy*. La Salle, IL: Open Court.

AESTHETICS

Abraham, Nicolas. 1986. "Psychoanalytic Esthetics: Time, Rhythm, and the Unconscious," *Diacritics*, vol. 16, no. 3 (Fall), pp. 3-14.

Bachelard, Gaston. 1964. *The Poetics of Space*. Mario Jolas, trans. Boston: Beacon Press.

Barthes, Roland. 1957. *Mythologies*. Annette Lavers, trans. London: Paladin.

Barthes, Roland. 1977. *Image, Music, Text*. Stephen Heath, trans. London: Fontana.

Beardsley, Monroe. 1966. *Aesthetics from Classical Greece to the Present: A Short History*. New York: Macmillan.

Brisson, Harriet E. 1992. "Visualization in Art and Science," *Leonardo*, vol. 25, nos. 3 and 4, pp. 257-262.

Bryson, Norman. 1983. *Vision and Painting: The Logic of the Gaze*. London: Macmillan.

Carroll, David. 1987. *Paraesthetics: Foucault, Lyotard, Derrida*. London: Methuen.

Chandrasekhar, Subrahmanyan. 1987. *Truth and Beauty: Aesthetics and Motivations in Science*. Chicago: University of Chicago Press.

Chordá, Frederic. 1991. "Computer Graphics for the Analysis of Perspective in Visual Art: `Las Meninas,' by Velázques," *Leonardo*, vol. 24, no. 5, pp. 563-567.

Gombrich, Ernst Hans. 1972. *Art, Perception, and Reality*. Baltimore: Johns Hopkins University Press.

Hume, Kathryn. 1984. *Fantasy and Mimesis Responses to Reality in Western Literature*. New York: Methuen.

Ivins, William. 1973. *On the Rationalization of Sight with an Examination of Three Renaissance Texts on Perspective*. New York: De Capo.

Kandinsky, Wassily. 1979. *Point and Line and Plane*. New York: Dover.

Kemp, Martin. 1990. *The Science of Art: Optical Themes in Western Art from Brunelleschi to Seurat*. New Haven, CT: Yale University Press.

Lefebvre, Henri. 1991. *The Production of Space*. Donald Nicholson-Smith, trans. Oxford: Blackwell.

Metz, Christian. 1974. *Film Language: A Semiotics of the Cinema*. Michael Taylor, trans. New York: Oxford University Press, 1974.

GENERAL

Bartholomew, C. 1979. *Mechanical Toys*. New York: Hamlyn.

Blonksy, Marshall, ed. 1985. *On Signs*. Oxford, UK: Basil Blackwell.

Castle, Terry. 1988. "Phantasmagoria: Spectral Technology and the Metaphorics of Modern Reverie," *Critical Inquiry*, vol. 15, no. 1 (Autumn), pp. 26-61.

Castoriadis, Cornelius. 1978. *History as Creation: Solidarity*. P. Cardan, trans. Pamphlet 54. London, July.

Castoriadis, Cornelius. 1984. *Crossroads in the Labyrinth*. K. Soper and M. Rhyle, trans. Sussex, UK: Harvester Press.

Castoriadis, Cornelius. 1987. *The Imaginary Institution of Society*. K. Blamey, trans. Cambridge: Polity Press.

Castoriadis, Cornelius. 1991. *Philosophy, Politics, Autonomy*. D. Curtis, trans. New York: Oxford University Press.

Cox, Donna J. 1992. "Caricature, Readymades and Metamorphosis: Visual Mathematics in the Context of Art," *Leonardo*, vol. 25, nos. 3 and 4, pp. 295-302.

Davis, Douglas. 1973. *Art and the Future: A History/Prophecy of the Collaboration between Science, Technology and Art*. London: Thames and Hudson.

Dickson, Stewart. 1992. "True 3D Computer Modelling: Sculpture of Numerical Abstraction," *Leonardo*, vol. 25, nos. 3 and 4, pp. 281-287.

Frisia, Emilio. 1992. "New Representative Methods for Real and Imaginary Environments," *Leonardo*, vol. 25, nos. 3 and 4, pp. 369-376.

Perry, Charles O. 1992. "On the Edge of Science: The Role of the Artist's Intuition in Science," *Leonardo*, vol. 25, nos. 3 and 4, pp. 249-252.

Saffaro, Lucio. 1992. "On Some New Platonic Forms," *Leonardo*, vol. 25, nos. 3 and 4, pp. 289-290.

RESEARCH IN DESIGN THEORY AND METHODOLOGY:
A Selective Overview of Recent Work in English

by GERHARD BANSE

INTRODUCTION: DESIGN SCIENCE—A NEW BRANCH OF SCIENCE

ACKNOWLEDGMENTS

PART I. CONFERENCES

1. Series of Conferences in the 1960s and 1970s
2. International Conference on Design Theory and Methodology (DTM)
3. International Conference on Engineering Design (ICED)

PART II. PERIODICALS

1. Journals
2. Others

PART III. BIBLIOGRAPHY

INTRODUCTION: DESIGN SCIENCE—A NEW BRANCH OF SCIENCE

However, what we have now is more like chaos than pluralism.

—Dixon (1987, p. 146)

In difference lies a source of strength. Long live the difference.

—Gregory (1979, p. 2)

The field of design concerns many disciplines: architecture, graphic art, interior decoration, industrial design, systems theory, and engineering. In recent years, researchers in engineering design have also made important contributions to the philosophy of technology. These and other scientists seek to understand, to describe, to explain, to prescribe, to predict, and to plan the process and result of the imaginary devising and outlining of novelty or innovation, the conditions of the possibility of "design." Because of such diversity, there is a broad and interesting discussion which contains different points of view, strategies of problem solving, and types of answers. Discussed themes include the meaning and different kinds of design (design as a process or a result, and design in art, in architecture, and in engineering); the history of design research; social and scientific needs for research in design; the scientific origins and rules of design; contributions from sociology, psychology, education or pedagogy, human biology, and logic; kinds of design knowledge, and more.

Questions at the most fundamental level are such as, "Is design theory more formalized or descriptive?" and "Is it possible to state a general theory of design?" Many philosophical problems exist within this broad field, among them: design as a problem-solving process, the role of values and evaluation in design, types of thinking in design, the relationships between logic (on a broad scale) and heuristics, the influence of language, visual thinking and nonverbal knowledge (and its representation—as in sketches, drawings, and models), and the teachable and learnable in design.

But many discussions stand alone, carried out in a mono-disciplinary way, without relation to the contributions being made in other disciplines. In order to make significant progress in design theory today, it is necessary to bring together all the different disciplines and the results of their research into design. To this end, I suggest a multidisciplinary approach. It is the aim of the following bibliography to offer a preliminary overview of the literature (books and articles), authors, periodicals, and scientific activities that are relevant for anyone who wishes to pursue such a multidisciplinary investigation.

The following survey consists of: (1) a list of important (regular) conferences and other scientific events; (2) a list of journals and other periodically published materials; and (3) a bibliography of English literature in the field of design theory. (Most entries are from the last 10 to 15 years, but some important publications from the 1960s and 1970s are also included.)

Note: These lists are by no means comprehensive. They merely represent what have been judged to be the most important publications, authors, and directions in research and activities concerning the field of design research.

ACKNOWLEDGMENTS

Special thanks to the former Director of the Science, Technology, and Society (STS) Program at Pennsylvania State University, Professor Carl Mitcham, for his help in the elaboration of this overview. Thanks also to Professor W. Ernst Eder, of the Department of Mechanical Engineering at the Royal Military College of Canada, and to Professor Richard Buchanan, of the Department of Design at Carnegie Mellon University, for helpful discussions.

PART I. CONFERENCES

1. Series of Conferences in the 1960s and 1970s

Held in London 1962 (Jones and Thornlay 1963); Birmingham, UK 1965 (Gregory 1966); Portsmouth, UK 1967 (Broadbent and Ward 1969); Cambridge, Massachusetts 1969 (Moore 1970); London 1973; New York 1974 (Spillers 1974); Berkeley, California 1975; Portsmouth, UK 1976 (Evans, Powell, and Talbot 1982); and Portsmouth, UK 1980 (Jaques and Powell 1981).

2. International Conference on Design Theory and Methodology (DTM)

The latest were held in Montreal 1989 (Elmaraghy, Seering, and Ullmann 1989); Chicago 1990 (Rinderle and Mellon 1990); Miami 1991 (Stauffer 1991); Scottsdale 1992 (Taylorn and Stauffer 1992); Albuquerque 1993 (Hight and Stauffer 1993); Minneapolis 1994 (Hight and Mistree 1994); and Boston 1995 (Ward 1995).

3. International Conference on Engineering Design (ICED)

Held in Rome 1981 (Hubka 1981a; Hubka 1981b; Hubka and Eder 1981); Copenhagen 1983 (Hubka 1983); Hamburg 1985 (Hubka 1985); Boston 1987 (Eder 1987a); Budapest 1988 (Hubka, Baratossy, and Pighini 1988); Harrogate 1989 (Hubka 1989); Dubrovnik 1990 (Hubka and Kostelic 1990); Zurich 1991 (Hubka 1991); The Haag 1993 (Roozenburg 1993); Prague 1995 (Hubka and Program Commitee 1995); and Tampere 1997. For the ICED in general, see also Sheldon 1994.

PART II. PERIODICALS

1. Journals

Design (London, UK)—monthly, from 1949 to 1994.

Design Book Review (Berkeley, CA)—quarterly, since 1983.

Design Issues: History, Theory, Criticism (Cambridge, MA)—three times annually, since 1984.

Design Management Journal (Boston, MA)—quarterly, since 1990.

Design Methods: Theories, Research, Education and Practice (San Luis Obispo, CA)—quarterly, since 1992.

Design Methods and Theories: Journal of the Design Methods Group (DMG) and Design Research Society (DRS) (Berkeley, CA)—quarterly, from 1976 to 1991; continued by *Design Methods: Theories, Research, Education and Practice* since 1992. This journal was titled the *DMG Newsletter* from 1966 through 1971, and the *DMG-DRS Journal: Design Research and Methods* from 1972 through 1975.

Design Quarterly (Minneapolis, MN)—quarterly, since 1946.

Design Studies: The International Journal for Design Research in Engineering, Architecture, Products, and Systems (Oxford, UK)—quarterly, since 1979. "It is an aim of *Design Studies* to enrich our understanding of design activities and to improve the practice of design" (Gregory 1979, p. 2). This journal was founded "with emphasis on design science and design in other fields" (Hubka and Eder 1996, p. 54).

International Journal of Technology and Design Education (Leeds, UK)—quarterly, since 1990.

Journal of Art and Design Education (Abingdon, UK)—three times annually, since 1982.

Journal of Design and Manufacturing: The International Research Journal of Concurrent Engineering (London, UK)—quarterly, since 1991. The scope of this journal is "the broad area of contemporary design and manufacturing methodologies....[It] had been created to occupy a unique position as perhaps the only journal reporting contemporary research activities which drive the development of design principles to capture to designer's intention for a product" (Nnaji 1991, p. i).

Journal of Engineering Design (Abingdon, UK)—quarterly, since 1990.

Research in Engineering Design: Theory, Applications, and Concurrent Engineering (New York)—quarterly, since 1989.

In my opinion, the most important of these journals in the field of theory or science of engineering design are *Design Studies*, *Journal of Engineering Design*, and *Research in Engineering Design*.

2. Others

(Proceedings of the) *International Conferences on Design Theory and Methodology* (Elmaraghy, Seering, and Ullmann 1989; Hight and Stauffer 1993; Hight and Mistree 1994; Rinderle and Mellon 1990; Stauffer 1991; Taylor and Stauffer 1992; Ward 1995).

WDK [Workshop-Design-Konstruktion] Series—Works from members of the International Society for the Science of Engineering Design (like the International Conference on Engineering Design (ICED); also Eder 1992, 1994; Hubka 1981; Roozenburg and Eekels 1990). These proceedings mostly feature such topics as "design science," "methodical design," "evaluation and decision in design," "theory of technical systems," and "design education."

BIBLIOGRAPHY

Abbott, Howard. 1980. *Safe Enough to Sell? Design and Product Liability.* London: Design Council. Pp. 192.

Abbott, Howard. 1997. *Safer by Design: A Guide to the Management and Law of Design for Product Safety.* London: Design Council. Pp. xviii, 329. First published 1987.

Adelson, Beth. 1989. "Cognitive Research: Uncovering How Designers Design; Cognitive Modeling: Explaining and Predicting How Designers Design," *Research in Engineering Design*, vol. 1, no. 1, pp. 35-42.

Aguirre, G.J., and Ken M. Wallace. 1990. "Evaluation of Technical Systems at the Design Stage," in Hubka and Kostelic, eds., 1990, vol. 1, pp. 414-423.

Albano, L.D., and N.P. Suh. 1992. "Axiomatic Approach to Structural Design," *Research in Engineering Design*, vol. 4, no. 3, pp. 171-183.

Alexander, Christopher. 1971. *Notes on the Synthesis of Form*. Cambridge, MA: Harvard University Press. Pp. 216. First published 1964.

Alexander, E.R. 1982. "Design in the Decision-Making Process," *Policy Sciences*, vol. 14, no. 3, pp. 279-292.

Altshuller, G.S. 1984. *Creativity as an Exact Science*. London: Gordon and Breach. Pp. xi, 319.

Andersson, S., and A.B.J. Hugnell. 1991. "Functional Modeling in Mechanical Engineering Design," in Hubka, ed., 1991, vol. 1, pp. 17-24.

Andreasen, M. Myrup. 1987. "Design Strategy," in Eder, ed., 1987a, vol. 1, pp. 171-178.

Andreasen, M. Myrup., Kahler, S., Lund, T., and K.G. Swift. 1988. *Design for Assembly.* 2nd ed. London: Springer.

Andreasen, M. Myrup, and J. Olesen. 1990. "The Concept of Dispositions," *Journal of Engineering Design*, vol. 1, no. 1, pp. 17-36.

Andreasen, M. Myrup. 1988. "Design Methodology in the Framework of Integrated Product Development," in Hubka, Baratossy, and Pighini, eds., 1988, vol. 1, pp. 24-30.

Andreasen, M. Myrup. 1991. "Design Methodology," *Journal of Engineering Design*, vol. 2, no. 4, pp. 321-335.

Anonymous. 1994. "Design Quo Vadis? Strategies and Tools for the Future of European Design: A Recommendation to the European Community," *Design Issues*, vol. 10, no. 3, pp. 77-83.

Archer, L. Bruce. 1965. *Systematic Method for Designers*. London: Council for Industrial Design. Pp. 39. See also William Pope, *Overview: L. Bruce Archer's Systematic Method for Designers*, Kent, OH: Kent State University School of Architecture of Environmental Design, 1972. Pp. 60.

Archer, L. Bruce. 1979. "Design as a Discipline: Whatever Became of Design Methodology?" *Design Studies*, vol. 1, no. 1, pp. 17-18.

Arciszewski, T. 1990. "Design Theory and Methodology in Eastern Europe," in Rinderle and Mellon, eds., 1990, pp. 209-217.

Arnheim, R. 1993. "Sketching and the Psychology of Design," *Design Issues*, vol. 9, no. 2, pp. 15-19.

Ashford, Frederick Charles. 1969. *The Aesthetics of Engineering Design*. London: Business Books. Pp. 129.

Ashwin, C. 1984. "Drawing, Design and Semiotics," *Design Issues*, vol. 1, no. 2, pp. 42-52.

Asimow, Morris. 1962. *Introduction to Design*. Englewood Cliffs, NJ: Prentice Hall. Pp. 135. Important early textbook.

Bailey, Robert Leo. 1978. *Disciplined Creativity for Engineers*. Ann Arbor, MI: Ann Arbor. Pp. xvii, 614.

Bayazit, N. 1993. "Designing: Design Knowledge; Design Research; Related Sciences," in de Vries, Cross, and Grant, eds., 1993, pp. 121-136.

Beheshti, R. 1993. "Design Decisions and Uncertainty," *Design Studies*, vol. 14, no. 1, pp. 85-95.

Beitz, W. 1994. "Design Science: The Need for a Scientific Basis for Engineering Design Methodology," *Journal of Engineering Design*, vol. 5, no. 2, pp. 129-133.

Bell, D.G., Dean, L.T., and P.D. Hauck. 1991. "Mathematical Foundations of Engineering Design Processes," in Stauffer, ed., 1991, pp. 181-189.

Blessing, L., Chakrabarti, A., and Ken Wallace. 1995. "A Design Research Methodology," in Hubka and Program Committee, eds., 1995, vol. 1, pp. 50-55.

Blockley, D.I. 1992. "Engineering from Reflective Practice," *Research in Engineering Design*, vol. 4, no. 1, pp. 13-22.

Bonsiepe, G. 1995. "The Chain of Innovation: Science, Technology, Design," *Design Issues*, vol. 11, no. 3, pp. 33-36.

Booker, Peter Jeffrey. 1979. *A History of Engineering Drawing*. London: Northgate. Pp. xvi, 239. The only history that exists on this topic.

Booker, Peter Jeffrey. 1962. *Principles and Precedents in Engineering Design*. London: Institution of Engineering Designers. Pp. 32.

Booker, Peter Jeffrey, ed. 1964. *Conference on the Teaching of Engineering Design*. London: Institution of Engineering Designers.

Booker, Peter Jeffrey, ed. 1966. *Teaching Engineering Design* (Proceedings of second conference). London: Institution of Engineering Designers.

Broadbent, Geoffrey, and Anthony Ward, eds. 1969. *Design Methods in Architecture*. London: Lord Humphries. Pp. 204.

Broadbent, Geoffrey. 1988. *Design in Architecture: Architecture of the Human Sciences*. London: D. Fulton. Pp. xviii, 524. First published 1973.

Broadbent, Geoffrey. 1979. "The Development of Design Methods: A Review," *Design Methods and Theories: Journal of the DMG*, vol. 13, no. 1, pp. 41-45.

Bucciarelli, Louis L. 1994. *Designing Engineers*. Cambridge, MA: MIT Press. Pp. x, 220. An ethnographic study arguing a moderate social constructivism.

Buchanan, Richard. 1992. "Wicked Problems in Design Thinking," *Design Issues*, vol. 8, no. 2, pp. 5-21. Reprinted, Margolin and Buchanan, eds., 1995.

Buck, Christopher Hearn. 1963. *Problems of Product Design and Development*. Oxford, UK: Pergamon. Pp. viii, 172.

Buhl, Harold R. 1960. *Creative Engineering Design*. Ames, IA: Iowa State University Press. Pp. 195.

Buur, J. 1990. "Design Methods in Japan: Research, Education, and Industrial Application from a European Viewpoint," in Hubka and Kostelic, eds., 1990, vol. 1, pp. 53-60.

Cain, William Dennis. 1969. *Engineering Product Design*. London: Business Books. Pp. xii, 278.

Carpenter, Stanley R. 1974. "Modes of Knowing and Technological Action," *Philosophy Today*, vol. 18, no. 2, pp. 162-168.

Carpenter, Stanley R. 1978. "The Cognitive Dimension of Technological Change," in *Research in Philosophy and Technology*, vol. 1, ed. Carl Mitcham. Stamford, CT: JAI Press, pp. 213-228.

Chakrabarti, A., and T. Bligh. 1994. "An Approach to Functional Synthesis of Solution in Mechanical Conceptual Design—Part I: Introduction on Knowledge Representation," *Research in Engineering Design*, vol. 6, no. 3, pp. 127-141.

Chandrasekaran, B. 1989. "A Framework for Design Problem-solving," *Research in Engineering Design*, vol. 1, no. 2, pp. 75-86.

Chang, S. 1990. "A Scientific Approach towards Developing an Engineering Design Theory," in Hubka and Kostelic, eds., 1990, vol. 1., pp. 61-67.

Christiaans, H.C.M., and K. Venselaar. 1990. "Knowledge as a Basis for Creativity in Design," in Hubka and Kostelic, eds., 1990, vol. 1, pp. 68-75.

Churchman, C. West. 1987. "The Philosophy of Design," in Nadler, ed., 1987, pp. 1-3.

Cook, H.E., and M.R. Gill. 1992. "On System Design," *Research in Engineering Design*, vol. 4, no. 4, pp. 215-226.

Corbett, J., M. Dooner, J. Meleka, and C. Pym, eds. 1991. *Design for Manufacture: Strategies, Principles and Techniques*. Reading, MA: Addison Wesley. Pp. ix, 357.

Corfield, K.G. 1979. *Product Design: A Report*. London: National Economic Development Office. Pp. vi, 94.

Coyne, Richard D., M.A. Rosenman, A.D. Radford, M. Balachandran, and J.S. Gero. 1990. *Knowledge-Based Design Systems*. Reading, MA: Addison Wesley. Pp. xi, 567.

Coyne, Richard D., and A. Snodgrass. 1991. "Is Designing Mysterious? Challenging the Dual Knowledge Thesis," *Design Studies*, vol. 12, no. 3, pp. 124-131.

Cross, Nigel. 1972. *Design Participation: Proceedings of the Design Research Society's Conference*, Manchester, UK, September 1971. London: Academy. Pp. 124.

Cross, Nigel. 1984. "Towards an Understanding of the Intrinsic Values of Design Education," *Design Studies*, vol. 5, no. 1, pp. 31-39.

Cross, Nigel, ed. 1984. *Developments in Design Methodology*. New York: Wiley. Pp. x, 357.

Cross, Nigel. 1993. "A History of Design Methodology," in de Vries, Cross, and Grant, eds., 1993, pp. 15-27.

Cross, Nigel. 1993. "Science and Design Methodology: A Review," *Research in Engineering Design*, vol. 5, no. 2, pp. 63-69.

Cross, Nigel. 1994. *Engineering Design Methods: Strategies for Product Design*. 2nd ed. New York: Wiley. Pp. x, 179.

Cross, Nigel., and N. Roozenburg. 1992. "Modeling the Design Process in Engineering and in Architecture," *Journal of Engineering Design*, vol. 3, no. 3, pp. 325-337.

Dasgupta, S. 1994. "Testing the Hypothesis Law of Design: The Case of the Britannia Bridge," *Research in Engineering Design*, vol. 6, no. 1, pp. 38-57.

DeNicola, Daniel. 1995. "Paradigms and Paraphernalia: On the Relation between Theory and Technology in Science," in J.C. Pitt, ed., *New Directions in the Philosophy of Technology*. Dortrecht, The Netherlands: Kluwer, pp. 85-93.

Dickson, K. 1990. "The Role of Design in the Innovation Process," *Journal of Engineering Design*, vol. 1, no. 3, pp. 269-278.

Dieter, George E. 1983. *Engineering Design: A Materials and Processing Approach*. London: McGraw-Hill. Pp. xiv, 592. Reprinted, New York: McGraw-Hill.

Dimarogonas, A.D. 1993. "On the Axiomatic Foundation of Design," in Hight and Stauffer, eds., 1993, pp. 253-258.

Dixon, John R. 1966. *Design Engineering: Inventiveness, Analysis and Decision Making*. New York: McGraw-Hill. Pp. vii, 354. Reprinted, Malaban, FL: R.E. Krieger.

Dixon, John R. 1987. "On Research Methodology Towards a Scientific Theory of Engineering Design," *Artificial Intelligence for Engineering Design, Analysis and Manufacturing*, vol. 1, no. 3, pp. 145-157.

Dixon, John R. 1991. "Engineering Design Science: The State of Education," *Mechanical Engineering*, vol. 113, no. 2, pp. 64-67.

Dixon, John R., and S. Finger. 1989. "The Proper Study of Mankind Is the Science of Design," *Research in Engineering Design*, vol. 1, no. 1, p. 1.

Doordan, Dennis P., ed. 1995. *Design History: An Anthology.* Cambridge, MA: MIT Press. Pp. xiv, 274.

Doppelt, F.F. 1987. "Introduction and Overview," in *System Design: Behavioral Perspectives on Designers, Tools, and Organizations*, eds. W.B. Rouse and K.R. Boff. New York: North-Holland, pp. 1-5.

Dowlen, C. 1996. "Developing the Creative Environment for Engineering Design," in Eder, ed., 1996, pp. 161-164.

Duffy, A.H.B., and M. Myrup Andreasen. 1995. "Enhancing the Evolution of Design Science," in Hubka and Program Committee, eds., 1995, vol. 1, pp. 29-35.

Dym, Clive L. 1994. *Engineering Design: A Synthesis of Views.* New York: Cambridge University Press. Pp. xviii, 205.

Edel, D. Henry, ed. 1967. *Introduction to Creative Design.* Englewood Cliffs, NJ: Prentice Hall. Pp. xiii, 237.

Eder, W. Ernst. 1985. "Design Education: Survey and Problems," in Hubka, ed., 1985, vol. 2, pp. 989-998a.

Eder, W. Ernst, ed. 1987a. *Proceedings of the 1987 International Conference on Engineering Design.* 1987 International Congress on Planning and Design Theory, Boston, MA, August 17-20, 1987. (*Schriftenreihe WDK*, vol. 13.) 2 vols. New York: American Society of Mechanical Engineers.

Eder, W. Ernst. 1987b. "Survey of Design Education," in Eder, ed., 1987a, vol. 2, pp. 881-887.

Eder, W. Ernst. 1990. "Engineering Design: A Perspective on U.K. and Swiss Developments," in Rinderle and Mellon, eds., 1990, pp. 225-234.

Eder, W. Ernst. 1990. "Design Science: Meta-Science to Engineering Design," in Rinderle and Mellon, eds., 1990, pp. 327-335.

Eder, W. Ernst. 1991. "Editorial," *Design Studies*, vol. 12, no. 4, pp. 194-196.

Eder, W. Ernst. 1993. "Science in Engineering, One Component of the Science of Engineering Design," in de Vries, Cross, and Grant, eds., 1993, pp. 137-164.

Eder, W. Ernst. 1994. "Developments in Education for Engineering Design: Some Results of Fifteen Years of Workshop Design-Construction Activity in the Context of Design Research," *Journal of Engineering Design*, vol. 5, no. 2, pp. 135-144.

Eder, W. Ernst. 1995. "An Appropriate Structure for Design Science," in Hubka and Program Committee, eds., 1995, vol. 1, pp. 3-8.

Eder, W. Ernst. 1995. "Engineering Design: Art, Science, and Relationships," *Design Studies*, vol. 16, no. 1, pp. 117-127.

Eder, W. Ernst, ed. 1996a. *Proceedings of the Workshop on Engineering Design and Creativity (EDC), 1996. (Schriftenreihe WDK,* vol. 24.) Zurich: Heurista.

Eder, W. Ernst. 1996b. "Design Education and Creativity," in Eder, ed., 1996a, pp. 125-132.

Eder, W. Ernst. 1996c. "Summary of General Discussion Group Reports," in Eder, ed., 1996a, pp. 119-124.

Eder, W. Ernst, and W. Gosling. 1965. *Mechanical System Design.* Oxford, UK: Pergamon. Pp. xiii, 226.

Eder, W. Ernst, Hubka, V., Melezinek, A., and S. Hosnedl, eds. 1992. *Proceedings of the International Conference on Engineering Design Education.* Reading, UK. *(Schriftenreihe WDK*, vol. 21.) Zurich: Heurista.

Eekels, Johannes. 1988. "On Evaluation and Decision in Design: An Analysis of the Problem Area," in Hubka, Baratossy, and Pighini, eds., 1988, vol. 1, pp. 56-61.

Eekels, Johannes. 1990. "Engineering Design and the Environmental Issue," in Hubka and Kostelic, eds., 1990, vol. 1, pp. 453-462.

Eekels, Johannes. 1994. "The Engineer as Designer and as a Morally Responsible Individual," *Journal of Engineering Design*, vol. 5, no. 1, pp. 7-23.

Eekels, Johannes. 1995. "Some Historical Remarks on the Philosophy of Making and Design," in Hubka and Program Committee, eds., 1995, vol. 1, pp. 36-43.

Eekels, Johannes. 1995. "Values, Objectivity, and Subjectivity in Science and Engineering," *Journal of Engineering Design*, vol. 6, no. 3, pp. 173-189.

Eekels, Johannes, and N.F.M. Roozenburg. 1991a. "A Methodical Comparison of the Structure of Scientific Research and Engineering Design: Their Similarities and Differences," in Hubka, ed., 1991, vol. 1, pp. 58-69.

Eekels, Johannes, and N.F.M. Roozenburg. 1991b. "A Methodological Comparison of the Structures of Scientific Research and Engineering Design: Their Similarities and Differences," *Design Studies*, vol. 12, no. 4, pp. 197-203.

Ehrlenspiel, K., and T. John. 1987. "Inventing by Design Methodology," in Eder, ed., 1987, vol. 1, pp. 29-38.

Ehrlenspiel, K. 1994. "Theory of Technical Systems," *Journal of Engineering Design*, vol. 5, no. 2, pp. 117-128.

Elmaraghy, W.E., Seering, W.P., and D.G. Ullman, eds. 1989. *Design Theory and Methodology, DTM '89.* Presented at the 1989 ASME Design Technical Conferences, 1st International Conference on Design Theory and Methodology, Montreal, Quebec, Canada, September 17-21, 1989. *(Design Engineering Division*, vol. 17.) New York: American Society of Mechanical Engineers.

Eppinger, S.D., Whitney, D.E., Smith, R.P., and D.A. Gebala. 1990. "Organizing the Tasks in Complex Design Projects," in Rinderle and Mellon, eds., 1990, vol. 1, pp. 39-46.

Ertas, Atila, and J.C. Jones. 1996. *The Engineering Design Process*. 2nd ed. New York: Wiley. Pp. x, 614. First published 1993.

Evans, Bernie J., Powell, James A., and R. Talbot, eds. 1982. *Changing Design*. New York: Wiley. Pp. viii, 379.

Fenves, S.J., and I.E. Grossmann. 1992. "An Interdisciplinary Course in Engineering Synthesis," *Research in Engineering Design*, vol. 3, no. 4, pp. 223-231.

Ferguson, Eugene S. 1992. *Engineering and the Mind's Eye*. Cambridge, MA: MIT Press. Pp. xiv, 241. Well-illustrated study of visual design thinking in engineering.

Ferguson, Eugene S. 1992. "Designing the World We Live In," *Research in Engineering Design*, vol. 4, no. 1, pp. 3-11.

Findeli, Alain. 1994. "Ethics, Aesthetics, and Design," *Design Issues*, vol. 10, no. 2, pp. 49-68.

Finger, Susan, and John R. Dixon. 1989a. "A Review of Research in Mechanical Engineering Design—Part I: Descriptive, Prescriptive, and Computer-Based Models of Design Processes," *Research in Engineering Design*, vol. 1, no. 1, pp. 51-67.

Finger, Susan, and John R. Dixon. 1989b. "A Review of Research in Mechanical Engineering Design—Part II: Representations, Analysis, and Design for the Life Cycle," *Research in Engineering Design*, vol. 1, no. 2, pp. 121-137.

Flurscheim, Charles. 1977. *Engineering Design Interfaces: A Management Philosophy*. London: Design Council. Pp. 138.

Fox, Richard L. 1971. *Optimization Methods for Engineering Design*. Reading, MA: Addison Wesley. Pp. xii, 270.

Frankenberger, E., and H. Birkhofer. 1996. "Creativity in Design Practice—Empirical Investigations in Industry," in Eder, ed., 1996, vol. 1, pp. 85-92.

French, Michael J. 1985. *Conceptual Design for Engineers*. 2nd ed. Berlin: Springer. Pp. ix, 226.

French, Michael J. 1991. "Towards a New Science of Functional Design," in Stauffer, ed., 1991, pp. 241-244.

French, Michael J. 1988. *Invention and Evolution: Design in Nature and Engineering*. New York: Cambridge University Press. Pp. xvi, 324.

French, Michael J., Chaplin, R.V., and P.M. Langdon. 1996. "A Creativity aid for Designers," in Eder, ed., 1996, pp. 107-111.

Frost, R.B. 1994. "A Suggested Taxonomy for Engineering Design," *Journal of Engineering Design*, vol. 5, no. 4, pp. 399-410.

Fulcher, A.J., and P. Hills. 1996. "Towards a Strategic Framework for Design Research," *Journal of Engineering Design*, vol. 7, no. 2, pp. 183-193.

Garratt, James. 1991. *Design and Technology.* New York: Cambridge University Press. Pp. 319. A community college general education textbook. Integrates aesthetics and engineering design.

Gasparski, Woiciech. 1984. *Understanding Design: The Praxiological-Systemic Perspective.* Seaside, CA: Intersystems. Pp. xii, 209.

Gasparski, Woiciech. 1995. "Design Methodology as Seen from Praxiological—Systemic Perspective," in Hubka and Program Committee, eds., 1995, vol. 1, pp. 23-28.

Gebahla, D.A., and N.P. Suh. 1992. "An Application of Axiomatic Design," *Research in Engineering Design*, vol. 3, no. 3, pp. 149-162.

Gibson, John E. 1968. *Introduction to Engineering Design.* New York: Holt, Rinehart, and Winston. Pp. ix, 214.

Gasparski, Woiciech. 1995. *Design and Systems: A General Applications Methodology.* New Brunswick, NJ: Transaction. Pp. xvi, 463.

Gille, Bernhard. 1986. "Essay on Technical Knowledge," in B. Gille, ed., *The History of Techniques*, vol. 2. New York: Gordon and Breach, pp. 1136-1185, 1238-1242, and 1338-1339.

Glegg, Gordon Lindsay. 1969. *The Design of Design.* Cambridge, UK: Cambridge University Press. Pp. 93.

Glegg, Gordon Lindsay. 1972. *The Selection of Design.* Cambridge, UK: Cambridge University Press. Pp. 83.

Glegg, Gordon Lindsay. 1973. *The Science of Design.* Cambridge, UK: Cambridge University Press. Pp. vii, 93.

Glegg, Gordon Lindsay. 1981. *The Development of Design.* Cambridge, UK: Cambridge University Press. Pp. 80.

Grant, D.P. 1979. "Design Methodology and Design Methods," *Design Methods and Theories: Journal of the DMG*, vol. 13, no. 1, pp. 46-47.

Green, Marc, ed. 1992. *Knowledge Aided Design.* London: Academic. Pp. xvii, 281.

Gregory, Sydney A., ed. 1966. *The Design Method.* New York: Plenium. Pp. xiii, 354.

Gregory, Sydney A. 1979. "Editorial: *Design Studies*—The New Capability," *Design Studies*, vol. 1, no. 1, p. 2.

Hales, Crispin. 1985. "Designer as Chameleon," *Design Studies*, vol. 6, no. 2, pp. 111-114.

Hales, Crispin. 1993. *Managing Engineering Design.* Essex, UK: Longman. Pp. xx, 212.

Hall, Arthur David. 1962. *A Methodology for Systems Engineering.* Princeton, NJ: Van Nostrand. Pp. 478.

Hanson, T.F. 1997. *Engineering Creativity.* 2nd ed. Newhall, CA: T.F. Hanson. Pp. viii, 128. First published 1987.

Hany, E.A., and E.F. Moritz. 1996. "Determinants and Correlates of Creative Problem Solving in Mechanical Engineering," in Eder, ed., 1996, pp. 73-77.

Hard, M. 1994. "Technology as Practice: Local and Global Closure Processes in Diesel-Engine Design," *Social Studies of Science*, vol. 24, no. 3, pp. 249-285.

Harrisberger, Lee. 1982. *Engineermanship: The Doing of Engineering Design.* 2nd ed. Monterey, CA: Brooks/Cole. Pp. vii, 148.

Hickman, Larry. 1995. "Techniques of Discovery: Broad and Narrow Characterizations of Technology," in J.C. Pitt, ed., *New Directions in the Philosophy of Technology.* Boston, MA: Kluwer, pp. 207-218.

Hight, T.K., Hornberger, L., and L.A. Sanchez. 1993. "Toward Critical Thinking Mechanical Designers," in Hight and Stauffer, eds., 1993, pp. 189-193.

Hight, T.K., and F. Mistree, eds. 1994. *Design Theory and Methodology, DTM '94.* Papers presented at the 1994 ASME Design Technical Conferences, 6th International Conference on Design Theory and Methodology, Minneapolis, Minnesota, September 11-14, 1994. (*Design Engineering Division,* vol. 68.) New York: American Society of Mechanical Engineers.

Hight, T.K., and L.A. Stauffer, eds. 1993. *Design Theory and Methodology, DTM '93.* Papers presented at the 1993 ASME Design Technical Conferences, 5th International Conference on Design Theory and Methodology, Albuquerque, New Mexico, September 19-22, 1993. (*Design Engineering Division,* vol. 53.) New York: American Society of Mechanical Engineers.

Hill, Percy H. 1970. *The Science of Engineering Design.* New York: Holt, Rinehart, and Winston. Pp. xi, 372.

Hollins, Bill, and Stuart Pugh. 1989. *Successful Product Design: What to Do and When.* London: Butterworth. Pp. xv, 167.

Holt, K. 1988. "Has the Designer Forgotten the User?" in Hubka, Baratossy, and Pighini, eds., 1988, vol. 1, pp. 95-102.

Hongo, K. 1981. "Present Situation of Study and Education of Design Engineering in Japan," in Hubka, ed., 1981, pp. 65-73.

Hongo, K., and A. Amirfazli. 1994. "Design Philosophy," *Journal of Engineering Design*, vol. 5, no. 2, pp. 93-102.

Hongo, K., and A. Amirfazli. 1995. "Scientific Prediction of Design Implications and Responsibilities of Design: A Speculative Approach to Design Research," in Hubka and Program Committee, eds., 1995, vol. 1, pp. 9-16.

Hongo, K., and N. Nakajima. 1991. "Relevant Features of the Decade 1981-1991 for Theories of Design in Japan," *Design Studies*, vol. 12, no. 4, pp. 209-213.

Hoover, S.P., Rinderle, J.R., and S. Finger. 1991. "Models and Abstractions in Design," in Hubka, ed., 1991, vol. 1, pp. 46-57.

Hosnedl, S. 1996. "Comments to Creativity in Design Education," in Eder, ed., 1996, pp. 153-155.

Hubka, Vladimir, ed. 1981a. *Proceedings of the International Conference on Design Methodology.* Rome, 1981. (*Schriftenreihe WDK*, vol. 5.) Zurich: Heurista.

Hubka, Vladimir, ed. 1981b. *Proceedings of the International Conference on Teaching Engineering Design*. Rome, 1981. (*Schriftenreihe WDK,* vol. 6.) Zurich: Heurista.

Hubka, Vladimir. 1982. *Principles of Engineering Design*. 1st English ed. trans. and ed. W.E. Eder. London: Butterworth.

Hubka, Vladimir, ed. 1983. *Proceedings of the 1983 International Conference on Engineering Design: CAD, Design Methods*. Copenhagen, 1983. (*Schriftenreihe WDK,* vol. 10.) 2 vols. Zurich: Heurista.

Hubka, Vladimir, ed. 1985. *Proceedings of the 1985 International Conference on Engineering Design: Theory and Practice of Engineering Design in International Comparison*. Hamburg, 1985. (*Schriftenreihe WDK*, vol. 12.) 2 vols. Zurich: Heurista.

Hubka, Vladimir, ed. 1989. *Proceedings of the 1989 International Conference on Engineering Design*. Harrogate, 1989. (*Schriftenreihe WDK*, vol. 18.) 2 vols. London: Heurista.

Hubka, Vladimir, ed. 1991. *Proceedings of the 1991 International Conference on Engineering Design*. Zurich, 1991. (*Schriftenreihe WDK,* vol. 20.) 2 vols. Zurich: Heurista.

Hubka, Vladimir. 1992. "Design for Quality and Design Methodology," *Journal of Engineering Design*, vol. 3, no. 1, pp. 5-15.

Hubka, Vladimir. 1995. "Specialized Design Science," in Hubka and Program Committee, eds., 1995, vol. 1, pp. 56-61.

Hubka, Vladimir. 1996. "Intuition and Feel for Design," in Eder, ed., 1996, pp. 65-72.

Hubka, Vladimir, and Program Committee, eds. 1995. *Proceedings of the 1995 International Conference on Engineering Design: 10th International Conference on Engineering Design*. Prague, 1995. (*Schriftenreihe WDK*, vol. 23.) 4 vols. Zurich: Heurista.

Hubka, Vladimir, Andreasen, M. Myrup, Eder, W. Ernst, Pighini, U., Schlesinger, A., and M. Wyss. 1981. *Terminology of the Science of Design Engineering (in Six Languages)*. (*Schriftenreihe WDK,* vol. 3.) Zurich: Heurista.

Hubka, Vladimir, Andreasen, M. Myrup, Eder, W. Ernst, and Peter J. Hills. 1988. *Practical Studies in Systematic Design*. London: Butterworths. Pp. xiv, 138.

Hubka, Vladimir, Baratossy, J., and U. Pighini, eds. 1988. *Proceedings of the 1988 International Conference on Engineering Design: 5th International Conference on Engineering Design*. Budapest, 1988. (*Schriftenreihe WDK*, vol. 16.) 3 vols. Zurich: Heurista.

Hubka, Vladimir, and W. Ernst Eder, eds. 1981. *Proceedings of the 1981 International Conference on Engineering Design: 1st International Conference on Engineering Design*. Rome, 1981. (*Schriftenreihe WDK,* vol. 7.) Zurich: Heurista.

Hubka, Vladimir, and W. Ernst Eder. 1987. "A Scientific Approach to Engineering Design," *Design Studies*, vol. 8, no. 3, pp. 123-137.

Hubka, Vladimir, and W. Ernst Eder. 1988. *Theory of Technical Systems: A Total Concept Theory for Engineering Design.* Berlin: Springer-Verlag. Pp. xiv, 275.

Hubka, Vladimir, and W. Ernst Eder. 1989. "Design Education and Design Science," in *The Place of Design in the Engineering School: Proceedings of the S. Neaman International Workshop, Haifa, Israel, June, 1989.* Haifa: S. Neaman Press, pp. 31-55.

Hubka, Vladimir, and W. Ernst Eder. 1990. "Design Knowledge: Theory in Support of Practice," *Journal of Engineering Design*, vol. 1, no. 1, pp. 97-108.

Hubka, Vladimir, and W. Ernst Eder. 1992. *Engineering Design.* Zurich: Heurista. This is a revised and expanded edition of *WDK 1: Principles of Engineering Design* (Zurich: Heurista, 1987).

Hubka, Vladimir, and W. Ernst Eder. 1996. *Design Science: Introduction to the Needs, Scope, and Organization of Engineering Design Knowledge.* London: Springer. Pp. xi, 251. Major systematic text. Complements Pahl and Beitz 1996; provides some historical backgorund.

Hubka, Vladimir, and A. Kostelic, eds. 1990. *Proceedings of the 1990 International Conference on Engineering Design: 7th International Conference on Engineering Design.* Dubrovnik, 1990. (*Schriftenreihe WDK*, vol. 19.) 4 vols. Zurich: Heurista.

Hubka, Vladimir, and J.W. Schregenberger. 1987. "Paths Towards Design Science," in Eder, ed., 1987, pp. 3-14.

Hundal, M.S. 1990. "Research in Design Theory and Methodology in West Germany," in Rinderle and Mellon, eds., 1990, pp. 235-238.

Hundal, M.S. 1991. "A Methodical Procedure for Search of Solutions from Function Structures," in Hubka, ed., 1991, vol. 1, pp. 9-16.

Hundal, M.S. 1996. "Engineering Design Creativity: Intuitive and Systematic Approaches," in Eder, ed., 1996, pp. 93-100.

Jansson, D.S., and R.G. Mauldin. 1988. "Conceptual Design: Model, Methodology, and Experiment," in Hubka, Baratossy, and Pighini, eds., 1988, vol. 1, pp. 118-125.

Jacques, Robin, and James A. Powell, eds. 1981. *Design—Science, Method: Proceedings of the 1980 Design Research Society Conference.* Guildford: Werbury. Pp. xii, 332.

Jaskola, Z. 1988. "'Constructing' Design of the Elements of Machines in Virtue of General Theory of Systems and Theory of 'Constructions' (Engineering Design)," in Hubka, Baratossy, and Pighini, eds., 1988, vol. 1, pp. 126-133.

Johnson, Ray C. 1980. *Optimum Design of Mechanical Elements.* 2nd ed. New York: Wiley. Pp. xviii, 519. First published 1961.

Jonas, W. 1993. Design as Problem-Solving? or: Here Is the Solution—What Was the Problem?" *Design Studies*, vol. 14, no. 2, pp. 157-170.

Jones, John Christopher. 1992. *Design Methods*. New York: Van Nostrand. Pp. lxiv, 407. Prefaces by C. Thomas Mitchell and Timothy Emly Jones. First published as *Design Methods: Seeds of Human Futures*, 1970.

Jones, John Christopher. 1984. *Essays in Design*. New York: Wiley. Pp. xiii, 335.

Jones, John Christopher, and D.G. Thornley, eds. 1963. *Conference on Design Methods*. Papers Presented at the Conference on Systematic and Intuitive Methods in Engineering, Industrial Design, Architecture and Communications, London, September 1962. Oxford, UK: Pergamon. Pp. xiii, 222.

Joseph, S. 1996. "Design Systems and Paradigms," *Design Studies*, vol. 17, no. 3, pp. 227-239.

Juvinall, Robert C., and K.M. Marshek. 1991. *Fundamentals of Machine Component Design*. 2nd ed. New York: Wiley. Pp. xxvi, 804.

Karandikar, H., and J. Shupe. 1995. "The Study of Design Theory and Methodology in Mechanical Engineering: A Commentary," in Ward, ed., 1995, vol. 2, pp. 361-370.

Kardos, G. 1988. "Fast for Systematic Design," in Hubka, Baratossy, and Pighini, eds., 1988, vol. 1, pp. 142-148.

Katz, Randy H. 1985. *Information Management for Engineering Design*. New York: Springer. Pp. viii, 93.

Kivenson, Gilbert. 1971. *Durability and Reliability in Engineering Design*. New York: Hayden. Pp. xiii, 186.

Koen, Billy Vaughn. 1984. "Toward a Definition of the Engineering Method," *Engineering Education*, vol. 74, no. 3, pp. 150-155.

Koen, Billy Vaughn. 1985. *Definition of the Engineering Method*. Washington, DC: American Society for Engineering Education. Pp. 74.

Koen, Billy Vaughn. 1987. "Generalization of the Engineering Method to the Universal Method," *Engineering Education*, vol. 77, no. 1, pp. 214-221.

Koen, Billy Vaughn. 1991. "The Engineering Method," in P.T. Durbin, ed., *Critical Perspectives on Nonacademic Science and Engineering*. Bethlehem, PA: Lehigh University Press, pp. 33-59.

Koopmann, P.J. Jr. 1995. "A Taxonomy of Decomposition Strategies Based on Structures, Behaviors, and Goals," in Ward, ed., 1995, vol. 2, pp. 611-618.

Korn, J. 1995. "Development of a Design Theory," in Hubka and Program Committee, eds., 1995, vol. 1, pp. 17-22.

Kostelic, A., and D. Marjanovic. 1995. "General Design Theories—Pro et Contra," in Hubka and Program Committee, eds., 1995, vol. 1, pp. 68-73.

Kota, S., and C.L. Lee. 1993. "General Framework for Configuration Design—Part 1: Methodology," *Journal of Engineering Design*, vol. 4, no. 4, pp. 277-289.

Kota, S., and A.C. Ward. 1990. "Functions, Structures, and Constraints in Conceptual Design," in Rinderle and Mellon, eds., 1990, pp. 239-250.

Krick, Edward V. 1969. *An Introduction to Engineering and Engineering Design*. 2nd ed. New York: Wiley. Pp. vii, 220. Reprinted, R.E. Krieger, 1976.

Kroes, Peter. 1995. "Technology and Science-Based Heuristics," in J.C. Pitt, ed., *New Directions in the Philosophy of Technology.* Boston: Kluwer, pp. 17-39.

Laudan, Rachel. 1984. "Cognitive Change in Technology and Science," in Rachel Laudan, ed., *The Nature of Technological Knowledge: Are Models of Scientific Change Relevant?* Boston: Kluwer, pp. 83-105.

Lawson, Bryan. 1984. "Cognitive Strategies in Architectural Design," in Cross, ed., 1984, pp. 85-104.

Lawson, Bryan. 1990. *How Designers Think: The Design Process Demystified.* 3rd rev. ed. Boston: Architectural Press. Pp. ix, 318. Design methods and experimental psychology for architectural, industrial, and interior designers.

Laymon, Ronald. 1985. "Idealizations and the Testing of Theories by Experimentation," in P. Achinstein and O. Hannaway, eds., *Observation, Experiment, and Hypothesis in Modern Physical Science.* Cambridge, MA: MIT Press, pp. 147-173.

Laymon, Ronald. 1989. "Applying Idealized Scientific Theories to Engineering," *Synthese*, vol. 81, no. 3, pp. 353-371.

Laymon, Ronald. 1995. "Idealizations, Externalities, and the Economic Analysis of Law," in J.C. Pitt, ed., *New Directions in the Philosophy of Technology.* Dortrecht: Kluwer, pp. 185-206.

Layton, Edwin T. Jr. 1984. "Science and Engineering Design," in *Bridge to the Future: A Centennial Celebration of the Brooklyn Bridge* (*Annals of the New York Academy of Science*, vol. 424), pp. 173-181.

Leech, D.J. 1972. *Management of Engineering Design.* New York: Wiley. Pp. 258.

Lent, Deane. 1970. *Analysis and Design of Mechanisms.* 2nd ed. Englewood Cliffs, NJ: Prentice Hall. Pp. xiii, 423. First published 1961. Reprinted, Salem, WI: Sheffield, 1993.

Lewin, D. 1979. "On the Place of Design in Engineering," *Design Studies*, vol. 1, no. 2, pp. 113-117.

Lewis, W.P., and A.E. Samuel. 1989. *Fundamentals of Engineering Design.* New York: Prentice Hall. Pp. xviii, 256.

Leyer, Albert. 1974. *Machine Design.* Glasgow, UK: Blackie. Pp. 115.

Liddament, T. 1996. "The Metamorphosis of the Design Vocabulary," *Design Studies*, vol. 17, no. 3, pp. 303-318.

Madanshetty, S.I. 1995. "Cognitive Basis for Conceptual Design," *Research in Engineering Design*, vol. 7, no. 7, pp. 232-240.

Maher, M.L., Balachandran, M.B., and D.M. Zhang. 1995. *Case-Based Reasoning in Design.* Mahwah, NJ: Lawrence Erlbaum.

Mann, R.W. 1992. "Engineering Design," in *McGraw-Hill Encyclopedia of Science & Technology*, vol. 6, 7th ed. New York: McGraw-Hill, pp. 392-399.

Margolin, Victor, ed. 1989. *Design Discourse: History, Theory, Criticism.* Chicago: University of Chicago Press. Pp. x, 291. Twenty papers from *Design Issues*, mainly on aesthetic and interpretative questions.

Margolin, Victor, and Richard Buchanan, eds. 1995. *The Idea of Design: A Design Issues Reader.* Cambridge, MA: MIT Press. Pp. xxii, 285. Twenty-three papers from the journal *Design Issues* on interpretative and cultural concerns.

Matousek, Robert. 1974. *Engineering Design: A Systematic Approach.* London: Blackie. Pp. viii, 264. First published 1963.

Mayall, W.H. 1967. *Industrial Design for Engineers.* London: Iliffe. Pp. 142.

McLean, G.F. 1993. "Integrating Ethics and Design," *IEEE—Technology and Society Magazine*, vol. 12, no. 3, pp. 19-30.

Meerkamm, H. 1994. "Design for X-A Core Area of Design Methodology," *Journal of Engineering Design*, vol. 5, no. 2, pp. 145-163.

Middendorf, William H. 1969. *Engineering Design.* Boston: Allyn and Bacon. Pp. xiii, 286.

Mills, J.J. 1992. "A Taxonomy of the Product Realization Process Environment," *Research in Engineering Design*, vol. 4, no. 4, pp. 203-213.

Mitcham, Carl. 1991. "Engineering as Productive Activity: Philosophical Remarks," in P.T. Durbin, ed., *Critical Perspectives on Nonacademic Science and Engineering*. Bethlehem, PA: Lehigh University Press, pp. 80-117.

Mitcham, Carl. 1994. "Engineering Design Research and Social Responsibility," in K.S. Shrader-Frechette, *Ethics of Scientific Research.* Lanham, MD: Rowman and Littlefield, pp. 153-223.

Mitcham, Carl. 1995. "Ethics into Design," in R. Buchanan and V. Margolin, eds., *Discovering Design: Explorations in Design Studies.* Chicago: University of Chicago Press, pp. 173-189.

Moore, Gary T., ed. 1970. *Emerging Methods in Environmental Design and Planning*. Cambridge, MA: MIT Press. Pp. xv, 409.

Morrison, Duncan. 1968. *Engineering Design: The Choice of Favourable Systems*. New York: McGraw-Hill. Pp. xii, 202.

Mueller, J. 1996. "Acceptance Barriers as Justified and Serious Defense Reactions of Creative Designers—Not Always only Ill Will, Thinking Inertia or Old Hat," in Eder, ed., 1996, pp. 79-84.

Mullins, S., and J.R. Rinderle. 1991. "Grammatical Approaches to Engineering Design—Part I: An Introduction and Commentary," *Research in Engineering Design*, vol. 2, no. 3, pp. 121-135.

Nadler, Gerald, ed. 1987. *1987 International Congress on Planning and Design Theory.* Boston, August 17-20, 1987—Plenary and Interdisciplinary Lectures. New York: American Society of Mechanical Engineers.

Newbury, D. 1996. "Knowledge and Research in Art and Design," *Design Studies*, vol. 17, no. 2, pp. 215-219.

Nezel, I. 1996. "Standpoint of a Psychologist to Education (Encouragement) of Creativity," in Eder, ed., 1996, pp. 49-53.

Nnaji, B.O. 1991. "Editorial," *Journal of Design and Manufacturing*, vol. 1, no. 1, pp. i-ii.

O'Leary, J. 1991. "Editorial: Wrestling with Design," *Mechanical Engineering*, vol. 113, no. 2, pp. 2.

O'Shaughnessy, K., and R.H. Sturges. 1992. "A Systematic Approach to Conceptual Engineering Design," in Taylor and Stauffer, eds., 1992, pp. 283-291.

Ottoson, S. 1996. "Boosting Creativity in Technical Development," in Eder, ed., 1996, pp. 35-39.

Pahl, Gerhard. 1996. "Transfer Ability as Educational Goal: Results of a Discourse at Ladenburg," in Eder, ed., 1996, pp. 133-138.

Pahl, Gerhard, and Wolfgang Beitz. 1996. *Engineering Design: A Systematic Approach*. 2nd rev. ed. K. Wallace, L. Blessing, and F. Bauert, trans. K. Wallace, ed. London: Springer. Pp. xxx, 544. First published 1977. This book is an international standard; different versions are available in six languages. Contains many detailed examples.

Papalambros, Panos Y., and D.J. Wilde. 1988. *Principles of Optimal Design: Modeling and Computation*. New York: Cambridge University Press. Pp. xxi, 416.

Pessen, D.W. 1990. "Simplicity in Mechanical Design," in Hubka and Kostelic. eds., 1990, vol. 1, pp. 136-145.

Peters, T.J. 1992. "Mechanical Design Heuristics to Reduce the Combinatorial Complexity for Feature Recognition," *Research in Engineering Design*, vol. 4, no. 4, pp. 195-201.

Petroski, Henry. 1989. *The Pencil: A History of Design and Circumstance*. New York: Knopf.

Petroski, Henry. 1991. "Human Error and the Case for Case Histories in Design," *Journal of Engineering Design*, vol. 2, no. 1, pp. 81-88.

Petroski, H. 1994. *Design Paradigms: Case Histories of Error and Judgment in Engineering*. New York: Cambridge University Press. Pp. xii, 209. Classical cases (Vitruvius and Galileo) followed by four of modern bridge building (Dee, Britannia, Brooklyn, and Tacoma Narrows).

Pile, John F. 1979. *Design: Purpose, Form and Meaning*. Amherst, MA: University of Massachusetts Press. Pp. xii, 203. Aesthetics oriented.

Pitts, G. 1973. *Techniques in Engineering Design*. New York: Wiley. Pp. 154.

Polanyi, Michael. 1956. "Pure and Applied Science and Their Appropriate Forms of Organization," *Dialectica*, vol. 10, no. 3, pp. 231-243.

Polanyi, Michael. 1958. *Personal Knowledge: Towards a Post-Critical Philosophy*. Chicago: University of Chicago Press. See especially chapter 4, "Skills".

Poolen, Lambert J. van. 1989. "A Philosophical Perspective on Technological Design," *International Journal of Applied Engineering Education*, vol. 5, no. 3, pp. 319-329.

Przystupa, E.W. 1990. "The Metaphors in Creative Designing," in Hubka and Kostelic, eds., 1990, vol. 1, pp. 146-151.

Pugh, Stuart. 1981. "Concept Selection—A Method that Works," in Hubka, ed., 1981, pp. 497-506. See also "Concept Selection—A Method that Works," in Roozenburg and Eekels, eds., 1990, pp. 73-82.

Pugh, Stuart. 1990. "Engineering Design—Unscrambling the Research Issues," *Journal of Engineering Design*, vol. 1, no. 1, pp. 65-72.

Pugh, Stuart. 1990. *Total Design: Integrated Methods for Successful Product Engineering*. Reading, MA: Addison Wesley. Pp. xiii, 278.

Radcliffe, D.F. 1990. "Models of Visual Thinking by Novice Designers," in Rinderle and Mellon, eds., 1990, pp. 145-152.

Ramirez, M.R. 1996. "A Semantic, Learning Theory of Engineering Design," in Eder, ed., 1996, pp. 139-144.

Ray, Martyn S. 1985. *Elements of Engineering Design: An Integrated Approach*. Englewood Cliffs, NJ: Prentice Hall. Pp. xi, 266.

Redmond, J. 1995. "Design Philosophy and Product Research," in Hubka and Program Committee, eds., 1995, vol. 2, pp. 430-435.

Reich, Y. 1995. "A Critical Review of General Design Theory," *Research in Engineering Design*, vol. 7, no. 7, pp. 1-18.

Reuleaux, F. 1894. *The Constructor: A Hand-Book of Machine Design*. H.H. Suplee, trans. Philadelphia, PA: H.H. Suplee. Pp. 312.

Rinderle, J.M. 1990. "Design Relations," in Rinderle and Mellon, eds., 1990, pp. 267-272.

Rinderle, J.R. 1991. "Grammatical Approaches to Engineering Design—Part II: Melding Configuration and Parametric Design Using Attribute Grammars," *Research in Engineering Design*, vol. 2, no. 3, pp. 137-146.

Rinderle, J., and C. Mellon, eds. 1990. *Design Theory and Methodology, DTM '90*. Papers presented at the 1990 ASME Design Technical Conferences, 2nd International Conference on Design Theory and Methodology, Chicago, Illinois, September 16-19, 1990. (*Design Engineering Division*, vol. 27.) New York: American Society of Mechanical Engineers.

Rittel, H. 1973. "The State of the Art in Design Methods," *Design Methods and Theories*, vol. 7, no, 2, pp. 143-147.

Rittel, H., and M. Webber. 1993. "Dilemmas in a General Theory of Planning," *Policy Sciences*, vol. 4, no. 2, pp. 155-169.

Roe, P.H., Soulis, G.N., and V.K. Handa. 1967. *The Discipline of Design*. Experimental ed. Boston: Allyn and Bacon. Pp. xix, 284.

Rohatynski, R. 1990. "Process of Technical Design in Operational Approach," in Hubka and Kostelic, eds., 1990, vol. 1, pp. 152-158

Roozenburg, N.F.M., ed. 1993. *Proceedings of ICED '93: 9th International Conference on Engineering Design*. August 17-19, 1993, Netherlands Congress Centre, The Hague. (*Schriftenreihe WDK*, vol. 22.) 3 vols. Zurich: Heurista.

Roozenburg, N.F.M., and N.G. Cross. 1991. "Models of the Design Process: Integrating Across the Disciplines," *Design Studies*, vol. 12, no. 4, pp. 215-220.

Roozenburg, N.F.M., and J. Eekels. 1995. *Product Design: Fundamentals and Methods*. New York: Wiley. Pp. xiii, 408. A widely used textbook.

Roozenburg, N.F.M., and J. Eekels, eds. 1990. *EVAD—Evaluation, Decision in Design*. Reading, UK. (*Schriftenreihe WDK*, vol. 17.) Zurich: Heurista.

Rosen, D.W., and T.J. Peters. 1996. "The Role of Typology in Engineering Design Research," *Research in Engineering Design*, vol. 8, no. 2, pp. 81-98.

Rouse, W.B., and K.R. Boff. 1987. "Workshop Themes and Issues: The Psychology of System Design," in W.B. Rouse and K.R. Boff, eds., *System Design: Behavioral Perspectives on Designers, Tools, and Organizations*. New York: North-Holland, pp. 7-17.

Rowe, Peter G. 1987. *Design Thinking*. Cambridge, MA: MIT Press. Pp. x, 229. Architecture oriented.

Ruiz, C., and F. Koenigsberger. 1970. *Design for Strength and Production*. London: Macmillan. Pp. x, 270.

Sandler, B.Z. 1990. "Method for Conceptual Design (Model of Intuitive Search for Solutions)," in Hubka and Kostelic, eds., 1990, vol. 1, pp. 172-186.

Sandler, B.Z. 1994. *Computer-Aided Creativity: A Guide for Engineers, Managers, Inventors*. New York: Van Nostrand. Pp. xiv, 368.

Sargent, P. 1994. "Design Science or Nonscience," *Design Studies*, vol. 15, no. 4, pp. 389-402.

Schmitt, G.N., and C.C. Chen. 1991. "Classes of Design—Classes of Methods—Classes of Tools," in Hubka, ed., 1991, vol. 1, pp. 402-409.

Schoen, Donald A. 1988. "Designing: Rules, Types and Worlds," *Design Studies*, vol. 9, no. 3, pp. 181-190.

Schoen, D.A. 1984. "Design: A Process of Enquiry, Experimentation and Research," *Design Studies*, vol. 5, no. 3, pp. 130-131.

Schoen, D.A. 1984. "Problems, Frames and Perspectives on Designing," *Design Studies*, vol. 5, no. 3, pp. 132-136.

Sekimoto, S., and M. Ukai. 1994. "A Study of Creative Design Based on the Axiomatic Design Theory," in Hight and Mistree, eds., 1994, pp. 71-77.

Sheldon, D.F. 1990. "Editorial Comment," *Journal of Engineering Design*, vol. 1, no. 1, p. 3.

Sheldon, D.F. 1994. "Editorial: Vladimir Hubka and the Workshop Design-Konstruktion School," *Journal of Engineering Design*, vol. 5, no. 2, pp. 91-92.

Shephard, M.S. 1990. "Idealization in Engineering Modeling and Design," *Research in Engineering Design*, vol. 1, no. 3/4, pp. 229-238.

Shigley, Joseph Edward, and Charles R. Mischke. 1963. *Mechanical Engineering Design*. New York: McGraw-Hill. Pp. 631.

Siddall, James N. 1967. *Mechanical Design Reference Sources*. Toronto: University of Toronto Press. Pp. 156.

Simon, Herbert A. 1988. "The Science of Design: Creating the Artifical," *Design Issues*, vol. 4, no. 1/2, pp. 67-82. Included in Herbert A. Simon, *The Sciences of the Artificial*, 3rd. ed. (Cambridge, MA: MIT Press, 1996).

Smith, Charles O. 1988. *Introduction to Reliability in Design*. Malabar, FL: R.E. Krieger. Pp. xiv, 269. First published 1976.

Spillers, William R., ed. 1974. *Basic Questions of Design Theory*. Proceedings of the Symposium on Basic Questions of Design Theory, Columbia University, May, 1974. Amsterdam: North-Holland. Pp. xi, 530.

Spillers, William R., and S.L. Newsome. 1993. "Engineering Design, Conceptual Design, and Design Theory: A Report," in de Vries, Cross, and Grant, eds., 1993, pp. 103-120.

Spotts, M.F., and T.E. Shoup. 1998. *Design of Machine Elements*. 7th ed. Upper Saddle River, NJ: Prentice Hall. Pp. xvi, 829. First published 1950.

Spunt, Leonard. 1971. *Optimum Structural Design*. Englewood Cliffs, NJ: Prentice Hall. Pp. xvi, 168.

Srinivasan, S., and R.H. Allen. 1989. "Partitioning and Guided Search: A Generalized Approach to Problem Solving in Preliminary Design," in Elmaraghy, Seering, and Ullman, eds., 1989, pp. 37-43.

Starkey, C.V. 1992. *Engineering Design Decisions*. London: E. Arnold. Pp. ix, 222.

Starr, Martin Kenneth. 1963. *Product Design and Decision Theory*. Englewood Cliffs, NJ: Prentice Hall. Pp. 120.

Stauffer, L.A., ed. 1991. *Design Theory and Methodology, DTM '91*. Papers presented at the 1991 ASME Design Technical Conferences, 3rd International Conference on Design Theory and Methodology, September 22-25, 1991, Miami, Florida. (*Design Engineering Division,* vol. 31.) New York: American Society of Mechanical Engineers.

Suh, Nam P. 1990. *The Principles of Design*. New York: Oxford University Press. Pp. xiv, 401.

Taguchi, Gen'ichi. 1986. *Introduction to Quality Engineering: Designing Quality into Products and Processes*. White Plains, NY: Unipub/Quality Resources. Pp. vi, 191.

Takeda, H., Hamada, S., Tomiyama, T., and H. Yoshikawa. 1990. "A Cognitive Approach to the Analysis of Design Processes," in Rinderle and Mellon, eds., 1990, pp. 153-160.

Taleb-Bendiab, A., Oh, V., and M. French. 1993. "Semantic Modeling of Engineering Design," *Journal of Engineering Design*, vol. 4, no. 1, pp. 41-54.

Taura, T. 1995. "Design Science for Functional Design Process Modeling," in Hubka and Program Committee, eds., 1995, vol. 2, pp. 456-464.

Taylor, A.J. 1993. "The Parallel Nature of Design," *Journal of Engineering Design*, vol. 4, no. 2, pp. 141-152.

Taylor, D.L., and L.A. Stauffer, eds. *Design Theory and Methodology, DTM '92*. Papers presented at the 1992 ASME Design Technical Conferences, 4th International Conference on Design Theory and Methodology, Scottsdale, Arizona, September 13-16, 1992. (*Design Engineering Division*, vol. 42.) New York: American Society of Mechanical Engineers.

Taylor, L.E., and M.R. Hendersen. 1994. "The Role of Features and Abstraction in Mechanical Design," in Hight and Mistree, eds., 1994, pp. 131-140.

Tjalve, Eskild. 1979. *Short Course in Industrial Design*. London: Newnes-Butterworths. Pp. 207.

Tjalve, Eskild, Andreasen, M. Myrup, and F.F. Schmidt. 1979. *Engineering Graphic Modeling: Workbook for Design Engineers*. London: Newnes-Butterworths. Pp. 119.

Tomiyama, T. 1995. "A Design Process Model that Unifies General Design Theory and Empirical Findings," in Ward, ed., 1995, vol. 2, pp. 329-340.

Trousse, B., and H. Christiaans. 1995. "Design as a Discursive Activity," in Hubka and Program Committee, eds., 1995, vol. 2, pp. 436-443.

Tuomaala, J. 1996. "Creative Engineering Design—Summary of a Book," in Eder, ed., 1996, pp. 23-33.

Tuttle, Stanley B. 1967. *Mechanisms for Engineering Design*. New York: Wiley. Pp. viii, 178.

Ullman, D.G. 1989. "A Taxonomy of Mechanical Design," in Elmaraghy, Seering, and Ullman, eds., 1989, pp. 23-36.

Ullmann, D.G. 1991. "The Status of Design Theory Research in the United States," in Hubka, ed., 1991, vol. 2, pp. 794-802.

Ullman, D.G. 1992. "A Taxonomy for Mechanical Design," *Research in Engineering Design*, vol. 3, no. 3, pp. 179-189.

Ullman, D.G. 1994. "Issues Critical to the Development of Design History, Design Rationale and Design Intent Systems," in Hight and Mistree, eds., 1994, pp. 249-258.

Ullman, D.G., and B. D'Ambrosio. 1995. "A Taxonomy for Classifying Engineering Decision Problems and Support Systems," in Ward, ed., 1995, vol. 2, pp. 627- 637.

U.S. National Research Council. 1991. *Improving Engineering Design: Designing for Competitive Advantage*. Washington, DC: National Academy Press. Pp. x, 107.

Vidosic, Joseph P. 1969. *Elements of Design Engineering*. New York: Wiley. Pp. xi, 330.

Vincenti, Walter G. 1990. *What Engineers Know and How They Know It: Analytical Studies from Aeronautical History*. Baltimore, MD: Johns Hopkins University Press. Pp. viii, 326. Insightful case studies of, for example, airfoil design, propeller testing, and the innovation of flush rivets.

Vries, Marc J. de. 1993. "Design Methodology and Relationships with Science: Indroduction," in de Vries, Cross, and Grant, eds., 1993, pp. 1-14.

Vries, Marc J. de, Cross, Nigel, and Donald P. Grant, eds. 1993. *Design Methodology and Relationships with Science*. Boston: Kluwer. Pp. vi, 327. Proceedings from a major NATO research conference.

Ward, A.C., ed. 1995. "9th International Conference on Design Theory and Methodology: Presented at the 1995 Design Engineering Technical Conferences," in *Proceedings of the 1995 Design Engineering Technical Conferences*, vol. 2 *(Design Engineering Division*, vol. 68). New York: American Society of Mechanical Engineers, pp. 289-661.

Warfield, John N. 1994. *A Science of Generic Design: Managing Complexity through Systems Design*. Ames, IA: Iowa State University Press. Pp. xxxix, 588. First published 1989.

Welch, R.V., and John R. Dixon. 1991. "Conceptual Design of Mechanical Systems," in Stauffer, ed., 1991, pp. 61-68.

Welch, R.V., and John R. Dixon. 1994. "Guiding Conceptual Design Through Behavioral Reasoning," *Research in Engineering Design*, vol. 6, no. 3, pp. 169-188.

Whitney, D.E. 1990. "Designing the Design Process," *Research in Engineering Design*, vol. 2, no. 1, pp. 3-13.

Whitney, P., ed. 1985. *Design in the Information Environment: How Computing is Changing the Problems, Processes, and Theories of Design*. Carbondale, IL: Southern Illinois University Press. Pp. 175.

Wilcox, Alan D. 1987. *Engineering Design: Project Guidelines*. Englewood Cliffs, NJ: Prentice Hall. Pp. xi, 163.

Wilde, Douglas J. 1978. *Globally Optimal Design*. New York: Wiley. Pp. xii, 288.

Willem, R.A. 1990. "Design and Science," *Design Studies*, vol. 11, no. 1, pp. 43-47.

Woodham, Johnathan M. 1983. *The Industrial Designer and the Public*. London: Pembridge Press. Pp. viii, 142.

Woodson, Thomas T. 1966. *Introduction to Engineering Design*. New York: McGraw-Hill. Pp. viii, 434. Early and influential textbook.

Yoshikawa, H. 1983. "Designers' Designing Models," in Hubka, ed., 1983, vol. 2, pp. 338-344.

Zeid, I., Rajidi, P., Berg, C., and Y. Yener. 1991. "Towards a Design Theory Based on Maintainance Knowledge," in Hubka, ed., 1991, vol. 1, pp. 753-758.

Zeng, Y., and G.D. Cheng. 1991. "On the Logic of Design," *Design Studies*, vol. 12, no. 3, pp. 137-141.

Zenisek, J. 1996. "Sources and Conditions for Engineering Design Creativity," in Eder, ed., 1996, pp. 41-47.

Zhu, J.W. 1990. "A New Creative Method-Division-Variation-Combination Creative Method and Its Application in Developing New Products," in Hubka and Kostelic, eds., 1990, vol. 1, pp. 205-211.

Zwicky, F. 1969. *Discovery, Invention, Research through the Morphological Approach*. New York: Macmillan. Pp. ix, 276.

AUTHOR INDEX

SUBJECT INDEX

Printed and bound by CPI Group (UK) Ltd, Croydon, CR0 4YY

16/09/2026

14963401-0001